Lecture Notes of the Institute for Computer Sciences, Social Informatics and Telecommunications Engineering   272

More information about this series at http://www.springer.com/series/8197

Trung Q. Duong · Nguyen-Son Vo ·
Van Ca Phan (Eds.)

Quality, Reliability, Security and Robustness in Heterogeneous Systems

14th EAI International Conference, Qshine 2018
Ho Chi Minh City, Vietnam, December 3–4, 2018
Proceedings

 Springer

Editors
Trung Q. Duong
Queen's University Belfast
Belfast, UK

Van Ca Phan
Ho Chi Minh City University of Technology
and Education
Ho Chi Minh City, Vietnam

Nguyen-Son Vo
Institute of Fundamental
and Applied Sciences
Duy Tan University
Ho Chi Minh City, Vietnam

ISSN 1867-8211   ISSN 1867-822X   (electronic)
Lecture Notes of the Institute for Computer Sciences, Social Informatics
and Telecommunications Engineering
ISBN 978-3-030-14412-8   ISBN 978-3-030-14413-5   (eBook)
https://doi.org/10.1007/978-3-030-14413-5

Library of Congress Control Number: 2019933405

This Springer imprint is published by the registered company Springer Nature Switzerland AG
The registered company address is: Gewerbestrasse 11, 6330 Cham, Switzerland

Preface

We are delighted to introduce the proceedings of the 2018 European Alliance for Innovation (EAI) International Conference on Heterogeneous Networking for Quality, Reliability, Security and Robustness (Qshine). This conference brought together researchers, developers, and practitioners from around the world who are leveraging and developing heterogeneous networks. The theme of Qshine 2018 was "Internet of Things (IoT) Technology Enabling the Fourth Industrial Revolution."

The technical program of Qshine 2018 consisted of 13 full papers in oral presentation sessions at the main conference tracks. The conference tracks were: Track 1, Security and Privacy; Track 2, Telecommunications Systems and Networks; and Track 3, Networks and Applications. Aside from the high-quality technical paper presentations, the technical program also featured two keynote speeches. The two keynote speakers were Dr. Hoa Le Minh, Northumbria University at Newcastle, UK, and Dr. Vien Ngo, Queen's University Belfast, UK.

Coordination with the steering chair, Prof. Imrich Chlamtac, was essential for the success of the conference; we sincerely appreciate the constant support and guidance. It was also a great pleasure to work with such an excellent Organizing Committee and we thank them for their hard work in organizing and supporting the conference. In particular, we thank the Technical Program Committee, led by our TPC co-chairs, Dr. Nguyen-Son Vo and Dr. Van Ca Phan, who completed the peer-review process of technical papers and compiled a high-quality technical program. We are also grateful to our conference manager, Andrea Piekova, for the support, and all the authors who submitted their papers to Qshine 2018.

We strongly believe that Qshine provides a good forum for all researchers, developers, and practitioners to discuss all science and technology aspects that are relevant to heterogeneous networks. We also expect that future Qshine events will be as successful and stimulating, as indicated by the contributions presented in this volume.

February 2019

Trung Q. Duong
Nguyen-Son Vo
Van Ca Phan

Organization

Steering Committee

Imrich Chlamtac   University of Trento, Italy

Organizing Committee

General Co-chairs

Trung Q. Duong	Queen's University Belfast, UK
Dung Van Do	HCMC University of Technology and Education, Vietnam

TPC Co-chairs

Nguyen-Son Vo	Duy Tan University, Vietnam
Van Ca Phan	HCMC University of Technology and Education, Vietnam

Sponsorship and Exhibit Chair

Hoa Le-Minh   Northumbria University, UK

Local Co-chairs

Giang Hieu Le	HCMC University of Technology and Education, Vietnam
Quoc An Hoang	HCMC University of Technology and Education, Vietnam
Tam Minh Nguyen	HCMC University of Technology and Education, Vietnam
Kien Chi Le	HCMC University of Technology and Education, Vietnam
Thin Ngoc Chau	HCMC University of Technology and Education, Vietnam

Workshops Chair

Daniel Benevides da Costa   Federal University of Ceará, Brazil

Publicity and Social Media Co-chairs

Antonino Masaracchia	University of Palermo, Italy
Son Ngoc Pham	HCMC University of Technology and Education, Vietnam

Publications Chair

Nguyen-Son Vo   Duy Tan University, Vietnam

Web Chair

Phuc Quang Truong   HCMC University of Technology and Education, Vietnam

Panels Co-chairs

Son Ngoc Pham	HCMC University of Technology and Education, Vietnam
Nhu Gia Nguyen	Duy Tan University, Vietnam
Berk Canberk	Istanbul Technical University, Turkey

Demos Co-chairs

Son Ngoc Truong	HCMC University of Technology and Education, Vietnam
Sang Quang Nguyen	Tan University, Vietnam
Zoran Hadzi-Velkov	Ss. Cyril and Methodius University, Republic of Macedonia

Tutorials Co-chairs

Hung Manh Nguyen	HCMC University of Technology and Education, Vietnam
Hung Viet Dang	Duy Tan University, Vietnam
Tuan Le	Middlesex University, UK
Marco Di Renzo	SUPELEC, and the University of Paris–Sud XI, Paris, France

Conference Manager

Andrea Piekova	European Alliance for Innovation (EAI)

Technical Program Committee

Bin Wu	Tianjin University, China
Van Ca Phan	Ho Chi Minh City University of Technology and Education, Vietnam
Chen Chen	Xidian University, China
Chen Ling	Dalian University of Technology, China
Chung Ho	Queen's University Belfast, UK
Chunsheng Zhu	University of British Columbia, Canada
Cong Trang Mai	Queen's University Belfast, UK
Dac-Binh Ha	Duy Tan University, Vietnam
Dang Viet Hung	Duy Tan University, Vietnam
Guangjie Han	Hohai University, China
Jialin Liu	Dalian University of Technology, China
Jian Fang	Dalian University of Technology, China
Junqing Zhang	Queen's University Belfast, UK
Khan Ferdous Wahid	Digital Security, Architecture and Innovation, Airbus, Germany
Lei Shu	Guangdong University of Petrochemical Technology, China
Lei Wang	Dalian University of Technology, China
Liang Sun	Dalian University of Technology, China
Long Nguyen	Queen's University Belfast, UK

Ming Zhu	Dalian University of Technology, China
Muhammad Azhar Iqbal	Capital University of Science and Technology, Pakistan
Naigao Jin	Dalian University of Technology, China
Nguyen Gia Nhu	Duy Tan University, Vietnam
Nguyen Gia Tri	Duy Tan University, Vietnam
Nguyen-Son Vo	Duy Tan University, Vietnam
Panlong Yang	University of Science and Technology of China, China
Pham Ngoc Son	Ho Chi Minh City University of Technology and Education, Vietnam
Phong Nguyen	Queen's University Belfast, UK
Qianzhen Sun	Dalian University of Technology, China
Sang Nguyen	Duy Tan University, Vietnam
Songtao Lu	Iowa State University, USA
Sunyoung Lee	Queen's University Belfast, UK
Tie Qiu	Dalian University of Technology, China
Tiep Hoang	Queen's University Belfast, UK
Toan Doan	Queen's University Belfast, UK
Tran Trung Duy	Posts and Telecommunications Institute of Technology, Vietnam
Wei Chen	China University of Mining and Technology, China
Wenbing Zhao	Cleveland State University, USA
Xuan-Kien Dang	Ho Chi Minh City University of Transport, Vietnam
Zhangbing Zhou	China University of Geosciences, China
Zhaolong Ning	Dalian University of Technology, China
Zhenquan Qin	Dalian University of Technology, China

Contents

Improving Privacy for GeoIP DNS Traffic

Lanlan Pan[1]([✉]), Xuebiao Yuchi[2], Xin Zhang[3], Anlei Hu[3],
and Jian Wang[1]

[1] Geely Automobile Research Institute, Zhejiang 315336, China
abbypan@gmail.com
[2] Chinese Academy of Sciences, Beijing 100190, China
[3] China Internet Network Information Center, Beijing 100190, China

Abstract. Many authoritative nameservers today support GeoIP feature. EDNS
Client Subnet (ECS) extension helps GeoIP authoritative nameserver to address
the public recursive resolver's proximity IP problem. However, ECS raises
some privacy concerns since recursive resolver leaks client subnet information
on the resolution path to the authoritative nameserver. In this paper we introduce
an EDNS ISP Location (EIL) extension, to make privacy improvement for
GeoIP DNS traffic while preserve the ECS optimization on the end-user experience, reduce response latency, and increase cache-hit rate. We analysis 910.9K
Chinese IPv4 CIDR/24 subnets, find that 479.9K TEL subnets, 234.0K UNI
subnets, and 66.3K MOB subnets can enable EIL to optimize DNS traffic.

Keywords: DNS · Privacy · GeoIP · Client subnet · ECS · EIL

1 Introduction

In order to bring the web content as close to the users as possible, many authoritative
nameservers support GeoIP feature, return different responses based on the perceived
geographical location of the resolvers' IP addresses [1–6].

As Fig. 1 shows, there are two critical factors that can affect the response accuracy
of authoritative nameserver:

(1) Proximity IP Problem: Is the resolver's IP address close enough to the client's IP
address?
(2) GeoIP Database Problem: Does the authoritative nameserver use an GeoIP
database with high quality?

Public recursive resolvers such as Google Public DNS and OpenDNS offer free
DNS resolution services for global users. These servers are not close enough to many
users since the public recursive service providers couldn't deploy servers among each
country and each ISP's network [7].

Therefore, public recursive resolvers face to serious proximity IP problem. To
counter this problem, Google proposes an EDNS Client Subnet (ECS) extension [8] to

---

This is an extended version of an earlier extended abstract presented at the International Conference
on Privacy, Security and Trust, 2018.

T. Q. Duong et al. (Eds.): Qshine 2018, LNICST 272, pp. 1–13, 2019.
https://doi.org/10.1007/978-3-030-14413-5_1

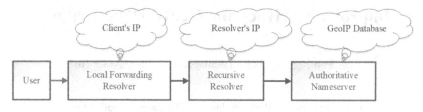

Fig. 1. GeoIP DNS traffic.

carry part of the client's IP address in the DNS packets for authoritative nameserver. As Fig. 2 shows, authoritative nameserver can directly use client subnet information to better understand where the end user is, while ignoring the resolver's IP address.

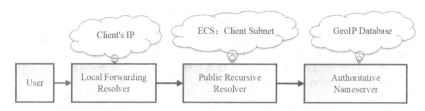

Fig. 2. ECS extension worked on GeoIP DNS traffic.

However, ECS also raises some privacy concerns because it leaks client subnet information on the resolution path to the authoritative nameserver. In [9], Kintis pointed out that ECS makes DNS communications less private: the potential for mass surveillance is greater, and stealthy, highly targeted DNS poisoning attacks become possible. Doileak [10] describe the privacy risk of ECS and why using a public DNS server might not improve your privacy.

To find the right balance between privacy improvement and end-user experience optimization, in this paper we introduce an EDNS ISP Location (EIL) extension. Compared with ECS, EIL can counter the proximity IP problem and GeoIP database problem more effectively, and improve privacy for GeoIP DNS Traffic.

The remainder of this paper is organized as follows. In Sect. 2, we discuss some related DNS privacy protection technologies. In Sect. 3, we describe the EIL extension in detail. From Sect. 4 to Sect. 6, we discuss response accuracy enhancement, privacy improvement, and operational benefit of EIL. In Sect. 7, we show our experiment on some GeoIP domains, and analyze the EIL effect. Finally, in Sect. 8, we discuss our work and conclude the paper.

2   DNS Privacy Protection Technologies

As Fig. 3 shows, most DNS privacy protection technologies [11, 12] can be divided into two groups. However, existing technologies are hard to provide user privacy controls on recursive resolvers that support ECS.

- Encrypting DNS Traffic

DNS over TLS [13], DNSCurve [14], DNSCrypt [15] and Confidential DNS [16] are different technologies to encrypt DNS traffic, they can improve the privacy on the resolution path, while none of them has any influence on the nameservers.

- Reducing Information Leakage to DNS Server

Root loopback [17] and QNAME minimization [18] can hide domain query information from root and TLD, while they are not designed for Client's IP privacy.

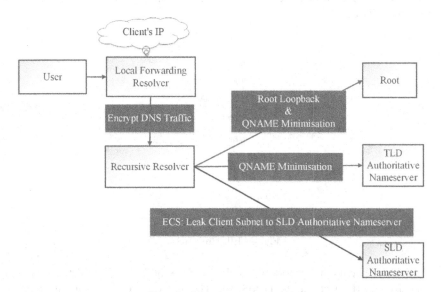

Fig. 3. DNS privacy protection technologies.

3   EDNS ISP Location (EIL) Extension

The EDNS ISP Location (EIL) extension proposed in this paper is similar to ECS. However, EIL includes the GeoIP information of client's IP in DNS packets, not client subnet information. Authoritative nameserver could provide a better answer by using GeoIP information of client's IP in EIL.

EIL can be added in DNS queries sent by recursive resolvers or local forwarding resolvers in a way that is transparent to stub resolvers and end users. EIL is only defined for the Internet (IN) DNS class.

3.1   Structure

EIL is structured as follows (Fig. 4):

- **OPTION-CODE**, 2 octets, defined in RFC6891 [19]. EDNS option code should be assigned by the IANA.

- **OPTION-LENGTH**, 2 octets, defined in RFC6891, contains the length of the payload (everything after OPTION-LENGTH) in octets.
- **COUNTRY**, 2 octets, uppercase, defined in ISO3166 [20], indicates the country information of the client's IP. For example, The COUNTRY of China is CN.
- **AREA**, 6 octets, uppercase, defined in ISO3166 country subdivision code, indicates the area information of the client's IP. For example, The AREA of Fujian Province in China is 35.
- **ISP**, 4 octets, uppercase, indicates the ISP information of the client's IP, using shortcut names. ISP shortcut names are unique within the context of the COUNTRY. As Table 1 shows, the shortcut name of China Telecommunications Corporation is TEL.

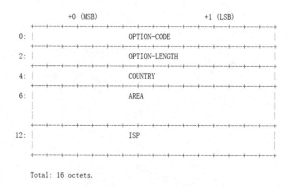

Total: 16 octets.

Fig. 4. EIL structure.

All fields of EIL are in network byte order. We use short names in the fields to limit the data size of EIL, decrease the DDoS risk. The null value 0×20 signifies that the field is unknown. If all fields in EIL are set to null value, means that client doesn't want to use EIL.

Table 1. China ISP.

ISP	ISP fullname
TEL	China Telecommunications Corporation
UNI	China United Network Communications
MOB	China Mobile Communications Corporation
TIE	China Tietong Telecommunications Corporation
EDU	China Education and Research Network

3.2 GeoIP Information

As Fig. 5 shows, Maxmind [21] gives the GeoIP information:

- Location: Quanzhou, Fujian, China, Asia
- ISP name: China Telecom

We can map Client's IP 61.154.123.91 into EIL <CN, 35, TEL>. Compared to ECS's client subnet such as 61.154.123.0/20, EIL contains very few sensitive information because it is associated with a very broad geographic area.

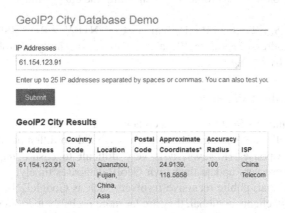

Fig. 5. Maxmind GeoIP information.

3.3 Deploy

Take Fig. 6 for example, when a public recursive resolver receives a DNS query from local forwarding resolver, it can map the client's IP to EIL <COUNTRY, AREA, ISP> information, then send EIL query to the authoritative nameserver. Using the GeoIP information specified in the EIL of DNS query, the authoritative nameserver can generate a tailored response.

Compared with ECS, EIL will move the GeoIP information mapping work from authoritative nameserver to recursive resolver, lighten the burden of authoritative nameserver, while it will increase DDoS risk on recursive resolver.

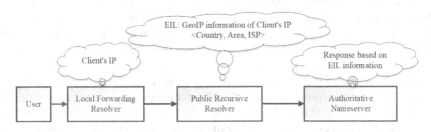

Fig. 6. EIL extension worked on GeoIP DNS traffic.

4 Response Accuracy Enhancement

4.1 Proximity IP Problem

ECS solves the proximity IP problem by generating the client subnet information from client's IP address.

Similar to ECS, EIL's GeoIP information <COUNTRY, AREA, ISP> is generated from client's IP address. Therefore, EIL also solves the proximity IP problem to GeoIP-enabled authoritative nameserver.

4.2 GeoIP Database Problem

GeoIP database quality affect response accuracy heavily. In ECS traffic mode, different GeoIP-enabled authoritative nameservers probably build up different GeoIP databases for their tailor response. However, it is very difficult to ensure huge amounts of authoritative nameservers update their own GeoIP databases timely.

On the other hand, public recursive resolvers such as GoogleDNS and OpenDNS serve magnanimity clients in global, they are far more probably to build up high quality GeoIP database than many small authoritative nameservers. Therefore, if public recursive resolvers such as GoogleDNS and OpenDNS support EIL, they can make sure huge amounts of authoritative nameservers return tailored response based on more precious GeoIP information, globally synchronized the enhancement of authoritative nameservers' response accuracy.

5 Privacy Improvement

5.1 Mitigating Client Subnet Leakage

The biggest privacy concern on ECS is that client subnet information is personally identifiable.

The more domains publish their zones on a third-party GeoIP-enabled authoritative nameserver, the more end user privacy information can be gathered by the third-party authoritative nameserver according to the ECS queries. Moreover, many authoritative nameservers only accept plaintext DNS queries, which means that the client subnet information is transparent on the resolution path from recursive resolver to authoritative nameserver.

EIL replaces the sensitive client subnet information to aerial view GeoIP information for user privacy protection. The GeoIP information is generated from Client's IP, not from user's physical geolocation. Even with EIL's most precise GeoIP information, authoritative nameserver can't identify sensitive personal information, and not any sensitive personal information is in plaintext DNS traffic from recursive resolver to authoritative nameserver. That is, EIL improves user privacy by sending less personal sensitive data than ECS.

5.2 Combating Targeted Censorship

DNS traffic is in plaintext by default. It is easily to be blocked or poisoned on internet. On plaintext mode, ECS query is fragile to targeted client subnet censorship.

However, since EIL's GeoIP information covers much bigger area than ECS's client subnet information, EIL will be stronger at monitoring targeted DNS censorship attack.

Encrypting the DNS traffic will be helpful to defense the targeted censorship in the future.

6 Operational Benefit

6.1 Cache-Hit Rate of Recursive Resolver

ECS sends the query with client subnet, which means that recursive resolvers send a new query to authoritative nameservers for each client subnet, even when they have known the response for some other GeoIP-closed client subnets. In fact, thousands of client subnets usually visit only a few target servers, there are many redundancy queries which can cause adverse effect on the average of response latency of recursive resolvers. Figure 7 shows a sample of ECS redundancy queries for www.qq.com.

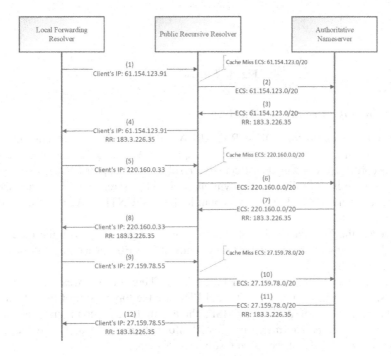

Fig. 7. ECS cache miss.

Each EIL GeoIP information covers huge amounts of client subnets. Therefore, compared to ECS redundancy queries to authoritative nameservers, EIL can sharply rise the cache-hit rate, reduce the response latency of recursive resolvers, lighten the burden of authoritative nameservers. Since EIL sends much less queries to authoritative nameservers, it can also improve privacy like qname minimization [18]. Figure 8 shows a sample of EIL optimized query for www.qq.com.

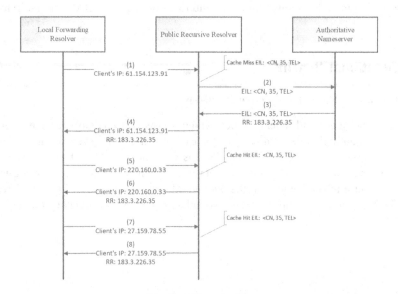

Fig. 8. EIL cache hit.

6.2 Cache Size of Recursive Resolver

EIL contains a whitelist for COUNTRY, AREA and ISP, which can be maintained by the IETF. Authoritative nameservers that support EIL can only response the EIL queries according to the whitelist. Recursive resolver that support EIL can only cache the EIL responses according to the whitelist too. Therefore, the EIL cache size of recursive resolver is related to the row count in the <COUNTRY, AREA, ISP> GeoIP information whitelist.

However, the ECS cache size of recursive resolver grows up with the number of client subnets. Obviously, under IPv6 environment, the EIL cache size will be much smaller than ECS.

Let's take the example of China in Table 2. There are 34 Areas in China. As Table 1 shows, TEL, UNI, MOB, TIE and EDU are the top 5 ISP in China. Consider about the null value of AREA and ISP, there will be 210 configurations on the authoritative nameserver to match the GeoIP information of China. This is the maximum cache size of EIL on recursive resolver for China.

Table 2. GeoIP information in China.

GEOIP TYPE	Configuration number
AREA + ISP	34 * 5 = 170
AREA + NULL ISP	34 * 1 = 34
NULL AREA + ISP	1 * 5 = 5
NULL AREA + NULL ISP	1 * 1 = 1
Total	170 + 34 + 5 + 1 = 210

6.3   DDoS Attack

To defense spoofed IP addresses, nameservers can optional implement EIL query only when the query is from a TCP connection. In case of pseudo-random sub-domain attack, nameservers may encounter error EIL queries padding with some random error string. As we have limited the data size of EIL, the defense cost will be smaller than sub-domain attack.

In strict defense mode, authoritative nameserver can refuse all error EIL query for security. However, for a better user experience, recursive resolver can also make a better EIL query instead of refusing if it thinks itself can afford.

7   Experiment

We describe the data collection and analysis of our study, which try to show the EIL major improvements on GeoIP DNS traffic described in previous sections. Our experiment code can be found in Github [22].

7.1   Data Collection

As Table 3 shows, we totally collect 910.9K Chinese IPv4 CIDR/24 subnets for experiment, which cover top 3 China ISPs and 31 Areas.

Table 3. Number of IPv4 CIDR/24 subnets from China Top 3 ISPs.

Country	ISP	Number of IPv4 CIDR/24 subnets
CN	TEL	497.8K
CN	UNI	272.3K
CN	MOB	140.8K

For each subnet, we send the ECS query for www.qq.com to authoritative name-server 123.151.66.83, get the tailored response, and add the GeoIP information.

Table 4 takes the client subnet 61.154.123.0/24 for example.

- We send the www.qq.com query to authoritative nameserver 123.151.66.83, with an ECS 61.154.123.0/24 option.

- 123.151.66.83 return the tailored response 183.3.226.35.
- We add the GeoIP information of response 183.3.226.35, which is <CN, 44, TEL>.
- We map the ECS 61.154.123.0/24 into EIL <CN, 35, TEL> GeoIP information.
- We finally build up the json-style response data for 61.154.123.0/24.

Table 4. Example ECS query and data collection.

Type	Value
Client subnet	61.154.123.0/24
ECS query	$ dig +short @123.151.66.83 www.qq.com +subnet = 61.154.123.0/24 www.qq.com 183.3.226.35

Response data	{
	domain: "www.qq.com",
	ecs_prefix: "61.154.123.0",
	ecs_mask: "24",
	eil_country: "CN",
	eil_area: "35",
	eil_isp: "TEL",
	response: [{
	ip: "183.3.226.35",
	ip_country: "CN",
	ip_area: "44",
	ip_isp: "TEL"
	}]
	}

7.2   Analysis

We count the subnets each response IP covered by <eil_country, eil_area, eil_isp> GeoIP information.

Table 5 shows <CN, 44, TEL> for example, the 44 means Guangdong area. We totally check 81720 subnets from <CN, 44, TEL>. The most frequent response IP is 183.3.226.35, which covers 80785 subnets, 98.85585%. Obviously, 183.3.226.35 virtually monopolize the <CN, 44, TEL> subnets' response, and the authoritative nameserver of www.qq.com supports GeoIP feature. If the authoritative nameserver and recursive resolver both enable EIL, then recursive resolver can directly return 183.3.226.35 as response to <CN, 44, TEL> subnets, avoid meaningless redundant ECS query traffic.

Table 6 shows <CN, 11, UNI> for example, the 31 means Beijing area. We totally check 32930 subnets from <CN, 11, UNI>. The top 2 frequent response IPs are 125.39.52.26 and 180.163.26.39, they cover 32520 subnets, 98.75493%. If the authoritative nameserver and recursive resolver both enable EIL, then recursive resolver can directly return 125.39.52.26 and 180.163.26.39 as response to <CN, 11, UNI> subnets, avoid meaningless redundant ECS query traffic.

Table 5. Response for <CN, 44, TEL>.

ID	Country	Area	ISP	Response IP	Subnets	Percent	Accumulate subnets	Accumulate percent
1	CN	44	TEL	183.3.226.35	80785	98.85585%	80785	98.85585%
2	CN	44	TEL	121.51.142.21	408	0.499266%	81193	99.35512%
3	CN	44	TEL	180.163.26.39	332	0.406265%	81525	99.76138%
4	CN	44	TEL	58.250.137.36	119	0.145619%	81644	99.90700%
5	CN	44	TEL	123.151.137.18	59	0.072198%	81703	99.97920%
6	CN	44	TEL	125.39.52.26	16	0.019579%	81719	99.99878%
7	CN	44	TEL	61.129.7.47	1	0.001224%	81720	100%

Table 6. Response for <CN, 11, UNI>.

ID	Country	Area	ISP	Response IP	Subnets	Percent	Accumulate Subnets	Accumulate Percent
1	CN	11	UNI	125.39.52.26	27289	82.86972%	27289	82.86972%
2	CN	11	UNI	180.163.26.39	5231	15.88521%	32520	98.75493%
3	CN	11	UNI	182.254.50.164	213	0.646827%	32733	99.40176%
4	CN	11	UNI	123.151.137.18	154	0.467659%	32887	99.86942%
5	CN	11	UNI	58.247.214.47	18	0.054661%	32905	99.92408%
6	CN	11	UNI	58.250.137.36	16	0.048588%	32921	99.97267%
7	CN	11	UNI	61.129.7.47	4	0.012147%	32925	99.98482%
8	CN	11	UNI	121.51.36.46	3	0.00911%	32928	99.99393%
9	CN	11	UNI	183.3.226.35	2	0.006073%	32930	100%

Table 7 shows <CN, 11, MOB> for example, the 31 means Beijing area. We totally check 53508 subnets from <CN, 11, MOB>. The top 3 frequent response IPs are 111.30.132.101, 121.51.142.21 and 121.51.36.46, they cover 51451 subnets, 96.15572%.

If we set the EIL enable threshold of the authoritative nameserver is top 2 frequent response IPs' accumulate percent is not less than 98.5%, authoritative nameserver can disable EIL response for <CN, 11, MOB> subnets, and the recursive resolver can send old ECS query traffic as before.

Table 7. Response for <CN, 11, MOB>.

ID	Country	Area	ISP	Response IP	Subnets	Percent	Accumulate subnets	Accumulate percent
1	CN	11	MOB	111.30.132.101	36041	67.35628%	36041	67.35628%
2	CN	11	MOB	121.51.142.21	8093	15.12484%	44134	82.48112%
3	CN	11	MOB	121.51.36.46	7317	13.67459%	51451	96.15572%
4	CN	11	MOB	111.30.144.71	2057	3.844285%	53508	100%

Table 8. EIL enable threshold of the authoritative nameserver.

Pseudocode
for each <COUNTRY, AREA, ISP> {
for my $id (1 .. $max_id){
if (Accumulate Percent >= $min_percent){
enable EIL;
set top $id Response IPs as EIL response.
}
}
}

Table 8 shows EIL enable threshold pseudocode of the authoritative nameserver.

For example, we can set $max\_id = 2$ and $min\_percent = 98.5\%$, Table 9 shows the EIL enable status. For TEL ISP, the authoritative nameserver of www.qq.com can enable EIL in 28 areas, which covered 479.9K subnets, 96.40245%. For UNI ISP, 26 areas can enable EIL, which covered 234.0K subnets, 85.92825%. For MOB ISP, 4 areas not enable EIL, the area codes are 11(Beijing), 32(Jiangsu), 31(Shanghai), 14 (Shanxi). We can find that responses for MOB ISP are not as steady as TEL ISP and UNI ISP, in this case, reserve ECS query can help for website traffic optimization.

Table 9. EIL enable decision for $max\_id = 2$, $min\_percent = 98.5\%$.

ISP	Enable EIL	Areas	Subnets	Percent
TEL	Yes	28	479.9K	96.40245%
	No	3	17.9K	3.597552%
UNI	Yes	26	234.0K	85.92825%
	No	5	38.3K	14.07175%
MOB	Yes	27	66.3K	47.10294%
	No	4	74.5K	52.89706%

8 Conclusion

We can't neglect the internet content delivery acceleration brought by ECS. The goal of EIL is to make privacy improvement for GeoIP DNS traffic while preserve the ECS optimization on the end-user experience, reduce response latency, and increase cache-hit rate.

We believe that EIL can provide user privacy controls both on public recursive resolvers and authoritative nameservers. Our future work is to do more experiments in China's network environment. We wish to apply the EIL into the real DNS traffic in the future, the IETF draft of EIL can be found in [23].

References

1. Amazon Route 53: Geolocation Routing. http://docs.aws.amazon.com/Route53/latest/ DeveloperGuide/routing-policy.html#routing-policy-geo
2. Using the GeoIP Features in BIND 9.10. https://kb.isc.org/article/AA-01149/0
3. DYN Predefined Geographic Groups of Traffic Director. https://help.dyn.com/traffic-director-predefined-geographic-regions/
4. Gdnsd Plugin Geoip. https://github.com/gdnsd/gdnsd/wiki/GdnsdPluginGeoip
5. PowerDNS GeoIP backend. https://doc.powerdns.com/md/authoritative/backend-geoip/
6. Microsoft Use DNS Policy for Geo-Location Based Traffic Management with Primary Servers. https://docs.microsoft.com/en-us/windows-server/networking/dns/deploy/primary-geo-location
7. Which CDNs support edns-client-subnet. http://www.cdnplanet.com/blog/which-cdns-support-edns-client-subnet/
8. Contavalli, C., van der Gaast, W., Lawrence, D., Kumari, W.: Client Subnet in DNS Queries. RFC7871 (2016)
9. Kintis, P., Nadji, Y., Dagon, D., Farrell, M., Antonakakis, M.: Understanding the privacy implications of ECS. In: Caballero, J., Zurutuza, U., Rodríguez, Ricardo J. (eds.) DIMVA 2016. LNCS, vol. 9721, pp. 343–353. Springer, Cham (2016). https://doi.org/10.1007/978-3-319-40667-1_17
10. The privacy risk of edns-subnet-client (ECS). https://www.doileak.com/blog-Public-DNS-might-not-%20improve-privacy.html
11. Bortzmeyer, S.: DNS privacy considerations. RFC 7626 (2015)
12. Grothoff, C., Wachs, M., Ermert, M., Appelbaum, J.: NSA's MORECOWBELL: Knell for DNS
13. Hu, Z., et al.: Specification for DNS over Transport Layer Security (TLS). RFC 7858 (2016)
14. Dempsky, M.: Dnscurve: link-level security for the domain name system. Work in Progress, draft-dempsky-dnscurve-01 (2010)
15. DNSCrypt. https://dnscrypt.org/
16. Wijngaards, W., Wiley, G.: Confidential DNS. IETF Draft (2015). https://tools.ietf.org/html/ draft-wijngaards-dnsop-confidentialdns-03
17. Kumari, W., Hoffman, P.: Decreasing Access Time to Root Servers by Running One on Loopback. RFC 7706 (2015)
18. Bortzmeyer, S.: DNS Query Name Minimisation to Improve Privacy. RFC7816 (2016)
19. Damas, J., Graff, M., Vixie, P.: Extension mechanisms for DNS (EDNS (0)). RFC 6891 (2013)
20. ISO 3166 Country Codes. http://www.iso.org/iso/country_codes
21. Maxmind GeoIP2 City Database. https://www.maxmind.com/en/geoip-demo
22. dns_test_eil. https://github.com/abbypan/dns_test_eil
23. Pan, L., Fu, Y.: ISP Location in DNS Queries. IETF Draft (2017). https://datatracker.ietf.org/ doc/draft-pan-dnsop-edns-isp-location/

Deep Reinforcement Learning Based QoS-Aware Routing in Knowledge-Defined Networking

Tran Anh Quang Pham[1(✉)], Yassine Hadjadj-Aoul[1],
and Abdelkader Outtagarts[2]

[1] Inria, Univ Rennes, CNRS, IRISA, Rennes, France
quang.pham-tran-anh@inria.fr, yassine.hadjadj-aoul@irisa.fr
[2] Nokia Bell Labs, Paris, France
abdelkader.outtagarts@nokia-bell-labs.com

Abstract. Knowledge-Defined networking (KDN) is a concept that relies on Software-Defined networking (SDN) and Machine Learning (ML) in order to operate and optimize data networks. Thanks to SDN, a centralized path calculation can be deployed, thus enhancing the network utilization as well as Quality of Services (QoS). QoS-aware routing problem is a high complexity problem, especially when there are multiple flows coexisting in the same network. Deep Reinforcement Learning (DRL) is an emerging technique that is able to cope with such complex problem. Recent studies confirm the ability of DRL in solving complex routing problems; however, its performance in the network with QoS-sensitive flows has not been addressed. In this paper, we exploit a DRL agent with convolutional neural networks in the context of KDN in order to enhance the performance of QoS-aware routing. The obtained results demonstrate that the proposed approach is able to improve the performance of routing configurations significantly even in complex networks.

Keywords: Knowledge-Defined networking ·
Software-Defined networking · Routing · Deep Reinforcement Learning

1 Introduction

Routing optimization and particularly traffic engineering are the most fundamental networking task and, therefore, has been extensively studied in a number of contexts. The emergence of Software-Defined networking (SDN) paradigm unveiled new capabilities in routing. In SDN, the control plane and the forwarding plane are separated. A centralized controller is responsible for computing routing decisions, thus reducing the complexity of network elements. Moreover, it is able to monitor the demands and availability of resources globally; therefore, it is capable of matching the resource needs optimally. However, the size of the solution space as well as the complexity of the optimization problem are

T. Q. Duong et al. (Eds.): Qshine 2018, LNICST 272, pp. 14–26, 2019.
https://doi.org/10.1007/978-3-030-14413-5_2

increased since SDN paradigm adds degree of freedoms in routing (flow-based forwarding vs destination-based forwarding) and the centralized controller has to solve the problem with a global view. As a result, new solutions of routing for SDN have been proposed [1,2]. In [3], the author proposed a Knowledge plane (KP), which is based on machine learning and cognitive techniques, to control the network. The KP is able to offer many advantages to networking, such as automation and recommendation and it may lead to a paradigm shift on the way we operate, manage, and optimize the data networks.

Machine learning (ML) techniques has been adopted and made breakthroughs in a number of application areas. ML algorithms can be classified into three categories: supervised learning (SL), unsupervised learning (USL), and reinforcement learning (RL). While SL and USL focuses on classification or regression tasks, RL algorithms learn to identify the best action series in order to maximize a given objective function (i.e reward). The most important advantages of ML is its capability of dealing with complicated problems; thus it is intuitive to exploit ML in the network domain where the complex problems are common [4]. In the context of routing, RL has been confirmed that it outperforms other ML techniques [5]; therefore, this paper focuses only on RL techniques.

RL techniques have been exploited to solve routing optimization [6]. They are also used in QoS routing [7]; however, table-based RL agents cannot provide efficient solutions for unseen network states. Deep Reinforcement Learning (DRL) is able to provide solutions for unseen network states, which cannot be achieved by traditional table-based RL agents. Moreover, it overcomes the iterative improvement process of optimization and heuristics by having a DRL agent providing a near-optimal solution in one single step. Recent breakthroughs in deep neural networks [8,9] has improved the performance of DRL; thus paving a way for adopting DRL in the context of networking. Deep convolutional neural networks [10] has proved its efficiency in various applications, e.g. image processing; therefore, the adoption of deep convolutional neural networks in the context of networking may offer some advantages.

In [5], the authors studied on the impacts of inputs and action spaces to the performance of machine learning in the context of routing problems. The results in the paper confirmed learning the link weights with traffic matrix outperforms other approaches. The link weights may have real positive values which cannot be solved by the well-known Deep Q Network algorithm [11] since it can handle only discrete and low-dimensional action space [12]. The authors in [12] addressed the needs of continuous control by proposing a Deep Deterministic Policy Gradient (DDPG) algorithm. The capability of DDPG in solving routing problem has been confirmed in [13]. However, the author in [13] focused on minimize the mean latency in the network and did not address the QoS routing problem with multiple metrics (e.g. latency and packet loss rate).

In this paper, we study on a DDPG agent which learns to make the QoS-aware routing decisions. Unlike to the DDPG agent presented in [13] which used fully connected layers, we take advantages of convolutional neural networks in order to extract the mutual impacts between flows in the networks; thus being able to provide better routing configurations. The rest of this paper is structured as follows. A brief review of KDN can be found in Sect. 2. A problem formulation followed by the proposed convolutional DRL network, which can improve the routing performance, is presented in Sect. 3. The performance of the proposed DRL network is verified by simulations, presented in Sect. 4.

2   Knowledge-Defined Networking

In the context of SDN architecture, a concept of Knowledge plane (KP) has been introduced in [3]. The addition of a KP to the conventional SDN paradigm unveils a new paradigm, what is called Knowledge-Defined Networking (KDN). In KDN, the data plane is responsible for storing, forwarding and processing data packets. Practically, it comprises line-rate programmable forwarding hardware which operate without being aware of the rest of the network and depend on the other planes to populate forwarding tables and update configurations.

The control plane exchanges operational states in order to inform the data plane about forwarding and processing rules. In SDN, a logically centralized SDN controller is responsible for this task. It programs the data plane via a southbound interface. The role of the management plane is to guarantee the proper operation and performance of the network in the long term. Its main tasks are to monitor the network and to provide network analytic. In fact, this task is could also be handled by the SDN controller. The KP exploits the control and the management planes to provide a comprehensive view and control over the network. Based on ML approaches, it is capable of learning the behavior of the network and operating the network appropriately.

The KDN paradigm uses a control loop to provide automation, optimization, validation, and estimation. Figure 1 describes the basic steps of the KDN control. The role of the analytics platform is to collect information to provide a completely global view of the network. It monitors the data plane elements in real-time and retrieves the control and management states from the SDN controller.

ML algorithms are able to learn the network behavior; thus, they play an important role in KP. The current and historical data offered by the analytics platform are fed to learning algorithms so as to generate knowledge. The decision maker (e.g. human decision or automatic decision) will give a decision based on knowledge and execute it via the northbound SDN controller API.

3   Convolutional Deep Learning for QoS-Routing

3.1   Network Model

We consider a framework in which a decision maker repeatedly selects routing configurations. Each flow is identified by the source and the destination. Traffic and QoS demands of each flow may be different. For instance, the video streaming may require a low latency and packet loss while web surfing is tolerant to packet

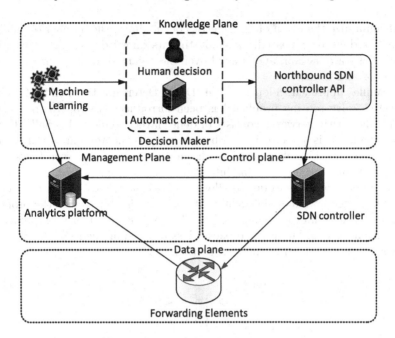

Fig. 1. KDN operation

loss and latency. Thus, we introduce a metric $\mathtt{QF}(\mathcal{R})$, named the number of qualified flows of routing configuration $\mathcal{R}$, which is the number of flows that meet their QoS requirements when applying the routing configuration $\mathcal{R}$.

We model the network as a capacitated indirected graph $\mathcal{G} = (\mathcal{V}, \mathcal{E}, c)$, where $\mathcal{V}$ and $\mathcal{E}$ are the vertex and edge sets, respectively, and $c : \mathcal{E} \leftarrow \mathbb{R}^+$ assigns a capacity to each edge. The number of nodes is $N = |\mathcal{V}|$ and the number of links is $L = |\mathcal{E}|$.

Traffic demands of flows are presented in a $N \times N$ matrix, named Traffic Matrix (TM), in which the entry (i, j) is the traffic demand between source i and destination j. Based on this matrix, the decision maker computes a L-vector, named link-weight vector, which is composed of weights of links in order to describe the routing configurations. This vector is exploited to calculate a forwarding table using the Dijkstra algorithm.

3.2 Deep RL Agent Model

The proposed RL agent is an off-policy, actor-critic, deterministic policy gradient algorithm [14] that interacts with the data network through states, actions, and rewards [15]. The state is the TM while the action is the link-weight vector. We define two reward functions: (1) the mean of QoS metrics (e.g. latency and end-to-end packet loss) and (2) the mean of qualified flows $\mathtt{QF}(\mathcal{R})$.

The objective of the agent is to identify the optimal policy π mapping from the state space S to the action space A, $\pi : S \rightarrow A$, that maximizes the expected reward (minimize the mean of QoS metrics or maximize the number of qualified flows). It is done by repeatedly ameliorating its knowledge of the connections between the state, action, and reward over the means of deep neural networks of the actor and the critic.

A vanilla approach adopting the Deep Deterministic Policy Gradient (DDPG) to solve routing problem has been introduced in [13]. In that paper, the actor and critic networks consists of dense layers connected serially. Consequently, the input is a vector in which each entry is the amount of a flow. This approach may not be able to explore the mutual impacts between flows. Consequently, we propose an extra module comprising multiple convolutional layers before the dense layers in order to allow the DRL network to learn the mutual impacts between flows more efficiently. Moreover, it is able to feed the proposed DRL network with multiple channels in order to provide a more comprehensive view of demands (e.g. latency and packet loss rate requirements). The details of this proposal is presented in the next section.

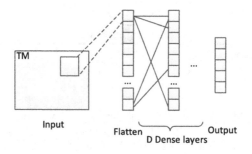

Fig. 2. Neural network architecture used in [13]

3.3   Convolutional DRL for QoS-Routing

Conventional neural network includes dense neural network layers as shown in Fig. 2. Unlike to fully connected layer in which each neuron connects to every neuron in the previous layer, a convolutional layer is only connected to a few local neurons in the previous layers; therefore it is relatively simpler than the fully connected layer. Moreover, it is able extract and learn mutual impacts of adjacent flows, thus enhancing the efficiency of routing configurations. Figure 3 presents the proposed neural network architecture.

TM is the input of the convolutional layers. There are M convolutional layers in critic network as well as in actor network. In fact, the value of M as well as the hyper-parameter of convolutional layers (e.g. stride size, filter size, etc.) may impact the DRL's performance. In the proposed architecture we may, also, have several max pooling layers. Having several layers of max pooling has the disadvantage of removing too much information, which significantly degrades the performance. However, a thorough study of these parameters is beyond the scope of our paper.

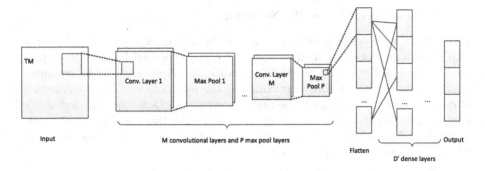

Fig. 3. Convolutional neural network architecture

We use the same value of M and hyper-parameters for both critic network and actor networks in all configurations in simulations. The output of M convolutional layers is flattned and fed into D' dense layer as the vanilla approach in order to compute the action (i.e. link-weight vector).

4   Simulation

To assess the performance of proposed approach we use a network topology BtEurope [16] of 24 nodes and 37 full-duplex links, with uniform link capacities (10 Mbps). The OMNeT++ discrete event simulator [17] (v5.4.1) was used to obtain the latency and packet loss rate under given traffic conditions (TM) and routing configurations ($\mathcal{R}$). We generate non-spare TM using a gravity model [18]. The total traffic entering (exiting) the network is generated by an exponential distribution with the mean 1000. The packet intervals of each flows follows an exponential distribution with the rate given in TM. The packet size is fixed at 1000 bits. We consider two QoS metrics: latency and packet loss rate. The QoS requirements of each flow is generated uniformly in range 1 ms to 100 ms for latency and 0 to 30% for packet loss rate.

Adam [19] is used for learning the neural network parameters. The learning rates for actor and critic networks are 10^{-4} and 10^{-3}, respectively. The discount factor γ is 0.99. The soft target is updated with the coefficient of $\tau = 0.001$. The batch size is 32. For the dense neural network learning, we reuse the configurations in [13], in which the number of dense layers is $D = 2$ with a numbers of units equal to 91 and 42, respectively. For the proposed approach, we used only one max pooling layer and 3 convolutional layers with 24 filters. The number of dense layers in the proposed approach is $D' = 1$, with 42 units.

We runs 10 simulations with different random seeds. The results are the mean of these runs. Each run is composed of 100 episodes, where one episode includes 50 different TMs.

4.1   Homogeneous Capacity Networks

The DDPG with dense neural networks is the approach used in [13] and the DDPG with convolutional neural networks is the proposed DRL network proposed in Sect. 3. For each type of DRL network, the objective could be maximizing the mean of QoS metrics or maximizing the mean of QFs. The combination of dense neural networks with the maximizing tmean QoS objective and maximizing mean of QFs objective are denoted as `Dense-QoS` and `Dense-QF`, while `Conv-QF` and `Conv-QoS` are the combination of convolutional neural networks with the QF objective and the mean QoS objective.

Figure 4 shows the mean latency under difference configurations of neural networks and objectives. The mean latency of `Dense-QoS` is slightly worse than of `Dense-QF`. It also applies to the packet loss rate and the mean number of QFs as shown in Figs. 5 and 6. It is because the QF objective aims to maximize the number of flows satisfying the QoS requirements and thus being able to distribute traffic better. As a result, the lower congestion level can be obtained and it leads to a better performance in the mean of QFs as well as the mean of QoS. Consequently, we focus on the mean QF objective in the following simulations.

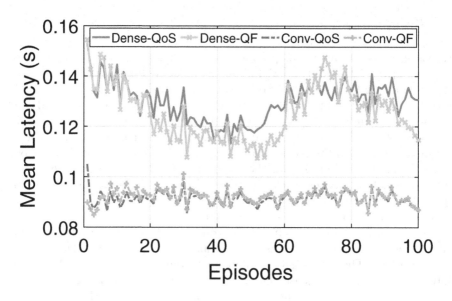

Fig. 4. Mean latency of dense vs convolutional networks

Generally, the performances of `Conv-QF` and `Conv-QoS` are better than of `Dense-QF` and of `Dense-QoS` in every QoS metric and the mean of QFs. It is because convolutional neural networks are able to extract the mutual impacts of flows; thus avoiding congestion better. A remarkable lower latency can be obtained by convolutional neural networks; however, the gaps in the mean of QFs are not always remarkable. It may be caused by the loose QoS requirements,

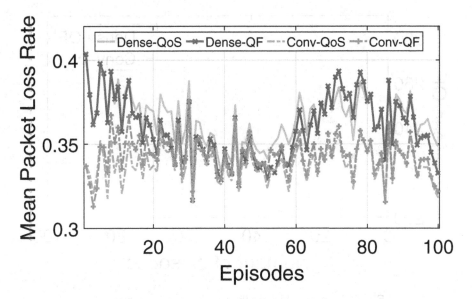

Fig. 5. Mean packet loss rate of dense vs convolutional networks

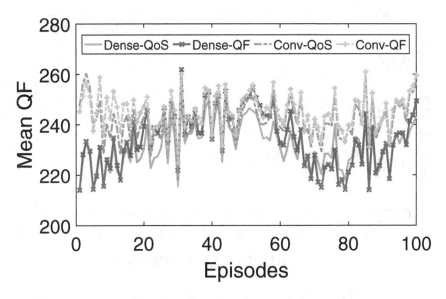

Fig. 6. Mean number of qualified streams of dense vs convolutional networks

especially the latency (up to 100 ms). To verify this hypothesis, we conduct the simulation with the latency in range of 1 ms to 10 ms and the loss rate in range of 0 to 10%. Figure 7 describes the mean of QFs in strict QoS requirements. Conv-QF outperforms Dense-QF. The gaps are significant.

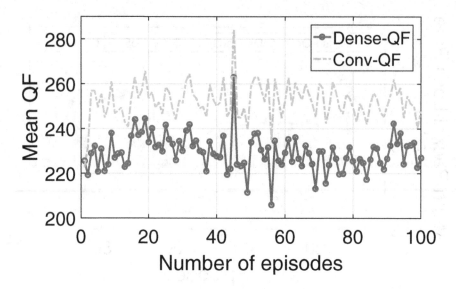

Fig. 7. Mean number of QF in strict QoS requirements

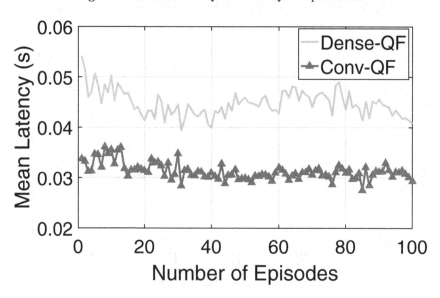

Fig. 8. Mean latency in heterogeneous networks

4.2 Heterogeneous Capacity Networks

In contrast to the previous section, the network composes of links with different capacities, i.e 10 Mbps, 20 Mbps, 50 Mbps, and 100 Mbps. This topology has higher capacity; however it is also more complicated than homogeneous networks.

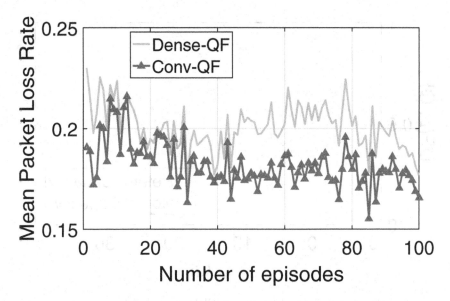

Fig. 9. Mean packet loss rate in heterogeneous networks

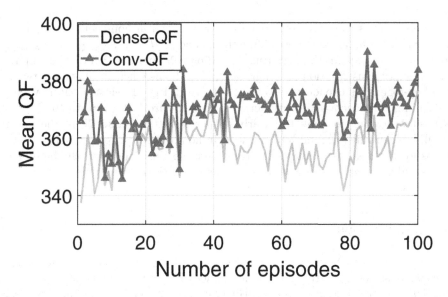

Fig. 10. Mean number of qualified streams under single and multiple channel input

The QoS requirements are from 1 ms to 100 ms for latency and 0 to 30% for packet loss rate.

Both latency and packet loss rate of Conv-QF are better than of Dense-QF as shown in Figs. 8 and 9. Consequently, Conv-QF has better the mean of QFs than Dense-QF as described in Fig. 10. The gaps between Conv-QF and Dense-QF

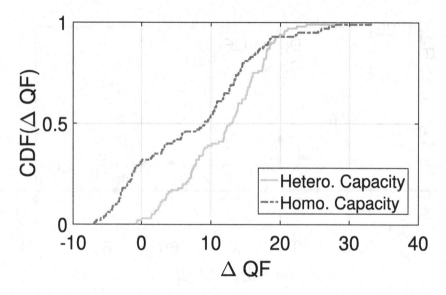

Fig. 11. Empirical CDF of Δ QF

in heterogeneous capacity networks are broader than in homogeneous capacity networks. To compare them, we define the metric $\Delta QF = QF_{\text{Conv-QF}} - QF_{\text{Dense-QF}}$ to indicate the gaps in the performance of Conv-QF and Dense-QF. The empirical cumulative distribution functions of ΔQF in homogeneous capacity network and heterogeneous capacity network are shown in Fig. 11. Most of ΔQF values (98%) in heterogeneous capacity network (capacity from 10 Mbps to 100 Mbps) is positive leading to Conv-QF clearly outperforms the Dense-QF in the mean of QFs. Meanwhile, around 30% of ΔQF in homogeneous capacity network (10 Mbps for all links) is negative. It means Conv-QF can cope with the heterogeneous and high capacity networks better than with the low and uniform capacity networks.

5   Conclusions

Knowledge Defined Networking is a potential paradigm for future data networks. This paper studied the structures of neural networks in deep reinforcement learning (DRL) in the context of routing. By adopting convolutional neural networks, the proposed mechanism is able to learn the mutual impacts between flows in the networks; therefore, it is able to provide a better routing configurations. The gaps is even broadened in heterogeneous capacity networks, indicating the advantages of convolutional neural networks in coping with complex scenarios. In future, we intend to extend this work by modifying DDPG algorithm in order to enhance the performance as well as apply it to more complicated problem in networking.

References

1. Akyildiz, I.F., Lee, A., Wang, P., Luo, M., Chou, W.: A roadmap for traffic engineering in SDN-OpenFlow networks. Comput. Netw. **71**, 1–30 (2014). https://doi.org/10.1016/j.comnet.2014.06.002, http://www.sciencedirect.com/science/article/pii/S1389128614002254
2. Layeghy, S., Pakzad, F., Portmann, M.: SCOR: software-defined constrained optimal routing platform for SDN. CoRR abs/1607.03243 (2016). http://arxiv.org/abs/1607.03243
3. Clark, D.D., Partridge, C., Ramming, J.C., Wroclawski, J.T.: A knowledge plane for the internet. In: Proceedings of the 2003 Conference on Applications, Technologies, Architectures, and Protocols for Computer Communications, pp. 3–10. ACM (2003)
4. Wang, M., Cui, Y., Wang, X., Xiao, S., Jiang, J.: Machine learning for networking: workflow, advances and opportunities. IEEE Netw. **32**(2), 92–99 (2018). https://doi.org/10.1109/MNET.2017.1700200
5. Valadarsky, A., Schapira, M., Shahaf, D., Tamar, A.: A machine learning approach to routing. CoRR abs/1708.03074 (2017). http://arxiv.org/abs/1708.03074
6. Boyan, J.A., Littman, M.L.: Packet routing in dynamically changing networks: a reinforcement learning approach. In: Advances in Neural Information Processing Systems, pp. 671–678 (1994)
7. Lin, S., Akyildiz, I.F., Wang, P., Luo, M.: QoS-aware adaptive routing in multilayer hierarchical software defined networks: a reinforcement learning approach. In: 2016 IEEE International Conference on Services Computing (SCC), pp. 25–33, June 2016. https://doi.org/10.1109/SCC.2016.12
8. Li, Y.: Deep reinforcement learning: an overview. arXiv preprint arXiv:1701.07274 (2017)
9. Arulkumaran, K., Deisenroth, M.P., Brundage, M., Bharath, A.A.: A brief survey of deep reinforcement learning. CoRR abs/1708.05866 (2017). http://arxiv.org/abs/1708.05866
10. Krizhevsky, A., Sutskever, I., Hinton, G.E.: Imagenet classification with deep convolutional neural networks. In: Advances in Neural Information Processing Systems, pp. 1097–1105 (2012)
11. Mnih, V., et al.: Human-level control through deep reinforcement learning. Nature **518**(7540), 529 (2015)
12. Lillicrap, T.P., et al.: Continuous control with deep reinforcement learning. arXiv preprint arXiv:1509.02971 (2015)
13. Stampa, G., Arias, M., Sanchez-Charles, D., Muntés-Mulero, V., Cabellos, A.: A deep-reinforcement learning approach for software-defined networking routing optimization. CoRR abs/1709.07080 (2017). http://arxiv.org/abs/1709.07080
14. Silver, D., Lever, G., Heess, N., Degris, T., Wierstra, D., Riedmiller, M.: Deterministic policy gradient algorithms. In: Proceedings of the 31st International Conference on International Conference on Machine Learning, ICML 2014, vol. 32, pp. I-387–I-395. JMLR.org (2014). http://dl.acm.org/citation.cfm?id=3044805.3044850
15. Sutton, R.S., Barto, A.G., et al.: Reinforcement Learning: An Introduction. MIT Press, Cambridge (1998)
16. The internet topology zoo. http://www.topology-zoo.org/dataset.html
17. Varga, A.: Discrete event simulation system. In: Proceedings of the European Simulation Multiconference (2011)

18. Roughan, M.: Simplifying the synthesis of internet traffic matrices. SIGCOMM Comput. Commun. Rev. **35**(5), 93–96 (2005). https://doi.org/10.1145/1096536. 1096551
19. Kingma, D.P., Ba, J.: Adam: a method for stochastic optimization. CoRR abs/1412.6980 (2014). http://arxiv.org/abs/1412.6980

Throughput Optimization for Multirate Multicasting Through Association Control in IEEE 802.11 WLAN

Dhrubajyoti Bhaumick$^{(\boxtimes)}$ and Sasthi C. Ghosh

Advanced Computing and Microelectronics Unit, Indian Statistical Institute,
203 B. T. Road, Kolkata 700108, West Bengal, India
dhrubabhaumick@gmail.com, sasthi@isical.ac.in

Abstract. Multicasting in wireless local area network is an efficient way
to deliver message from a source user to a specified group of destination users simultaneously. In unirate multicasting, all users belonging
to a particular group receive their services at the same basic rate. This
may underutilize network resources as users requirements are generally
heterogeneous in nature. To resolve this limitation, multirate multicasting is introduced, where different users belonging to a particular group
may receive their services at different rates. Often dense deployment of
access points (APs) is required for coverage and capacity improvement.
Thus an station (STA) may come under the coverage range of several APs
and hence there may exists many possible associations between the STAs
and the APs. Hence finding an efficient association is very important as
individual throughput of the STAs as well as the overall system throughput depend on it. We have developed an efficient algorithm to find an
appropriate association for multirate multicasting. The objective is to
maximize overall system throughput while respecting the user fairness.
Through simulations, we have evaluated and compared the performance
of our proposed algorithm with other well-known metrics such as received
signal strength indicator, minimum hop-distance, in-range STA number
and normalized cost. Results show that the proposed algorithm significantly improves the overall system throughput in comparison to these
metrics.

Keywords: IEEE 802.11 WLAN · Multirate multicast ·
Multicast association · Association control ·
Throughput maximization · User fairness

1 Introduction

Multicasting is a technique in which a message or information can be delivered
from a source user to a group of destination users simultaneously. Recently multicasting for multimedia applications like live lectures, online examinations, and
video conferences are increasingly being used in several sectors. The IEEE 802.11

© ICST Institute for Computer Sciences, Social Informatics and Telecommunications Engineering 2019
Published by Springer Nature Switzerland AG 2019. All Rights Reserved
T. Q. Duong et al. (Eds.): Qshine 2018, LNICST 272, pp. 27–47, 2019.
https://doi.org/10.1007/978-3-030-14413-5_3

wireless local area network (WLAN) has become the most popular and widely used wireless Internet access technology because of its low-cost and high-speed connectivity to the users. In an infrastructure based WLAN a set of access points (APs) is directly or via multi-hop connected to the Internet through a wired backbone network. A set of stations (STAs) access this network through these APs. An AP establishes a cell and coordinates all the communications that take place within that cell's area. Typically a dense deployment of the APs is required for coverage and capacity improvement in WLAN. Since many APs are deployed in a region, it is possible that an STA may be in the range of several APs, though it should be associated with only one AP at a time. In such situation, a well-known problem is which AP an STA selects to associate with [5,14]. An STA can receive data frames from an AP only when it is associated with that AP. This requires an STA to AP *association*. An STA may request different kind of services from the service provider such as *unicast services* and *multicast services*. When an STA maintains an association with an AP to get its unicast services then that kind of association is known as *unicast association*. Similarly, an association that is maintained by an STA with an AP to get its multicast services is known as *multicast association*. In this paper we deal with the multicast association problem and hence, in the rest of the paper, by association we mean to say multicast association.

There are two types of multicasting namely *unirate multicasting* and *multirate multicasting*. In an unirate multicasting, all the users belonging to a particular multicast group receive their services at the *same data rate* known as multicast session rate. This rate is determined according to the *worst channel condition* observed among all the users in a multicast group. Multicasting at this lower rate will occupy the transmission channel for a longer time and hence may *underutilize* the network resources [25]. In addition, it is not possible to meet the *heterogeneity* in user requirements by unirate multicasting. Hence it is desirable to use the multirate transmission to meet the heterogeneity in user requirements as well as efficient utilization of the network resources. In multirate multicasting, different users belonging to a particular group may receive their multicast services at different rates [12,13,21,22]. This does not imply separate rate for each user as multicasting is not a mere combination of several unicasting sessions. An AP must transmit its multicast packets to its associated STAs at the same rate. But different APs may transmit at different rates. The rate at which an AP will transmit multicast packets must be determined by taking into account the worst channel condition observed between the concerned AP and its associated STAs [3]. Thus unlike unirate multicasting there is no unique multicast session rate in multirate multicasting.

Time is divided into cycles, where a cycle duration is shared by both the unicast and multicast sessions. At the beginning of a multicast session, the APs and the STAs need to switch from their unicast to multicast mode within a fixed time interval. Such time-synchronization with respect to both APs and STAs can be achieved by network time protocol (NTP) [16,18]. Since multicast transmission at lower rate always results in the longer transmission time, designing a multicast algorithm that can reduce the channel occupancy time by increasing the transmission rate is desirable. This higher transmission rate will increase the overall system throughput by reducing the required number of time slots for completing the ongoing multicast session. This in effect helps to increase the unicast session duration as well.

In this paper, our main objective is to find an optimal association between the STAs and the APs such that the overall system throughput is maximized while taking care of the fairness of individual throughput obtained by the STAs. It is important to note that the problem of finding optimally fair utility allocation vector for multirate multicasting is NP-hard [21,23]. Therefore, we have developed a greedy algorithm for finding such optimal association which works as follows. If we associate an STA with an AP it may *pull-up* or *pull-down* the overall system throughput depending on the position of the concerned STA with respect to the positions of other STAs already associated with the concerned AP. An STA is associated with an AP based on the amount of throughput it pulls up or pulls down. The strategy is fair as an STA makes its association decision by considering not only its own throughput but also the throughput obtained by other STAs. We have compared the performance of the proposed algorithm with other well-known metrics like received signal strength indicator (RSSI) [4,5,7,8,14–16], minimum hop-distance [7,8,15,16], in-range STA number [7,8,15,16] and normalized cost [7,8,15]. Simulation results show that the proposed algorithm achieves much improved overall system throughput in comparison to these metrics.

The rest of the paper is organized as follows: Sect. 2 summarizes the related works. The system model is described in Sect. 3. Section 4 presents the problem statement and its mathematical formulation. The key idea of the solution approach is demonstrated through some motivational examples in Sect. 5. The proposed greedy algorithm is presented in Sect. 6. The time complexity of the proposed greedy algorithm is also presented in this section. The simulation results are presented in Sect. 7. Finally, Sect. 8 concludes the paper.

2   Related Works

The multicast association control in WLAN have been studied by several researchers. In [6], the authors have proposed an association strategy for supporting real-time multicast services in WLAN. In [15,16], the authors have proposed an association control mechanism for WLAN which optimizes the overall network load. An association strategy is proposed in [7,8] which maximizes the system throughput by controlling the multicast session data rate. All these studies are, however, based on the unirate multicasting.

Several authors have studied different aspects of multirate multicasting in WLAN. In [10], the authors have considered multirate transmissions to reduce the multicast/broadcast latency. An utility based multirate transmission is proposed in [12] which takes into account the heterogeneity in user requirement as well as the user fairness. A routing metric for reliable multicast in multirate WLAN environment have been studied in [20,25]. In [20], a routing and congestion control mechanism is proposed for multirate multicasting. A fair distributed congestion control mechanism for multirate multicast is presented in [22]. The problem of congestion control in networks which support the multirate multicasting have been studied in [11]. A low-overhead rate control and fair allocation of utilities for multirate multicasting is studied in [13,21]. Authors in [21,23] have shown that the problem of finding lexicographically optimal utility allocation vector for multirate multicasting is NP-hard. In [1,9,24] authors have considered the resource allocation problem for multicast services in WLAN. In [19], authors have shown that though the multirate multicasting improves the user quality of service (QoS), it also complicates the network optimization. They introduced a control scheme which dynamically optimizes the multirate multicast transmissions. A multirate multicasting method over wireless networks with time varying channel conditions and limited bandwidth have been proposed in [2], which dynamically adapts the transmission rate and forward error correction (FEC) for multicasting video traffic. In [17], a joint dynamic rate allocation and transmission scheduling optimization scheme based on opportunistic routing and network coding is proposed for scalable video multirate multicasting.

Though different aspects of multirate multicasting have been considered by several researchers, the multicast association problem in combination with the maximization of overall system throughput while respecting the user fairness has not been adequately studied. In this paper, we have developed an efficient greedy algorithm for finding an optimal association between the subscribed STAs and the available APs for multirate multicasting in WLAN which maximizes the overall system throughput while respecting the fairness of the individual throughput obtained by the users.

3   Network Model

The network model used in our study is described as follows. We have considered an infrastructure based WLAN where n number of APs are directly or via multi-hop connected through a wired backbone network to the main access point (MAP). The MAP is nothing but a special AP which has the backbone Internet connection. There are m number of subscribed STAs, which access this network through these APs. An AP establishes a cell and coordinates all the communications that take place within that cell's area. Typically an area is covered by multiple APs. Thus an STA may be in the coverage range of several APs. An STA can be associated with at most one AP at a time but an AP may serve multiple STAs simultaneously. We assume that an AP can serve at most 32 STAs simultaneously [15,16,29]. An STA can send/receive data packets via an AP only when it is associated to that AP. Network time is divided into cycles, where a cycle duration is shared by both the unicast and multicast sessions. The cycle duration as well as the unicast and multicast session intervals are configured by the network provider. The network service provider advertises these system information by means of beacon signals. At the beginning of a multicast session, the APs and the STAs need to switch from their unicast mode to the multicast mode at a fixed time interval. Such time-synchronization with respect to both APs and STAs may be achieved by network time protocol (NTP) [18].

Though according to the current IEEE 802.11 standard [26], multicast packets are transmitted to all the subscribed STAs at the same *basic data rate*, the feasibility of transmitting multicast packets at a rate higher than the basic rate have been established and studied by several authors [10,12,16]. According to IEEE 802.11 standard, IEEE 802.11b WLAN supports 1.0, 2.0, 5.5 and 11.0 Mbps data rates [26]. It uses dynamic rate shifting which allows the data rates to be automatically adjusted with the changing nature of the radio channel condition. For each rate, there is an optimal range for the successful operation at that rate. Most IEEE 802.11b vendors provide the optimal range for each data rate [27,28,30–32] in accordance with their supplied devices. Note that the optimal ranges vary with different vendors [27,28,30–32] and also with different models of the same vendor [31,32]. In our model, given the positions of the APs and the STAs, the physical rate at which an STA can be associated to an AP is determined based on these optimal ranges. The rate at which an AP transmits its multicast packets to its associated STAs and the overall system throughput obtained by an association are then computed based on these physical data rates. It is important to note that in our study we have considered the legacy of IEEE 802.11b standard for simplicity. However, our approach can be extended to any other standards (e.g., IEEE 802.11 a, g, n) as long as the operating rates and their respective optimal ranges are known.

4   Problem Statement and Its Mathematical Formulation

In order to get the multicast services, an STA must be associated with an AP at a certain physical rate. Let r_{ij} be this physical rate at which STA i can be associated with AP j. The value of r_{ij} can be computed based on the optimal ranges for different rates as stated earlier. We assume that STA i can be associated to AP j only if $r_{ij} \geq \tau$, where τ is a predefined threshold data rate. This threshold is determined based on the minimum rate required by an STA for decoding its multicast packets at the handset. This implies that an STA i will be considered as outside the coverage range of the network and can not be associated to any AP if $r_{ij} < \tau \; \forall \, j \in S_{AP}$, where S_{AP} is the set of available APs in the network.

In multirate multicasting, the rate at which an AP transmits its multicast packets is determined based on the worst channel condition observed between the AP and its associated STAs. All the associated STAs of an AP receive their multicast packets at this rate. Let r_j^{min} be this rate at which AP j transmits its multicast packets to its associated STAs. Hence to ensure the full coverage, the r_j^{min} of AP j must be set to the lowest of the data rates obtained by the STAs associated with AP j. That is $r_j^{min} = \min_i \{r_{ij}: $ STA i is associated to AP $j\}$. The throughput provided by AP j to its associated STAs can be represented as $\sigma_j = (r_j^{min} \times m_j)$, where m_j is the number of STAs associated with AP j. Here σ_j implies the total amount of data received by all the STAs associated with AP j per unit time. The overall system throughput is then defined as $\sigma = \sum_{j=1}^{n} \sigma_j$, where n is the total number of APs. Here σ represents the total amount of data received by all the STAs per unit time.

Now, from the network service provider prospectives, it is very important to increase the overall system throughput as their gross revenue is directly depending on it. Let C be the cost per unit data usage. Then the gross revenue earned or generated by the Internet service provider from the current multicast session is $G_{revenue} = C \times \sum_{j=1}^{n} \sigma_j$. It is evident from the said equation that the gross revenue earned by the network service provider is directly proportional with the overall system throughput. It is clear that if we are able to achieve higher value of overall system throughput for ongoing multicast session which in tern generate more amount of gross revenue for the network service provider. So, in this paper our main objective is to find an association between the APs and the STAs such that the overall system throughput is maximized.

Let S_{AP} be the set of APs and S_{STA} be the set of STAs which receive at least τ data rate from at least one AP in S_{AP}. Let $\mathscr{C}$ be the set of available data rates. It is important to note that r_j^{min} $(j \in S_{AP})$ is a real variable belongs to $\mathscr{C}$, where $\mathscr{C} = \{1.0, 2.0, 5.5, 11.0\}$. For all $i \in S_{STA}$ and $j \in S_{AP}$, we define the following binary variables.

$$x_{ij} = \begin{cases} 1 & \text{if STA } i \text{ is associated with AP } j \\ 0 & \text{otherwise.} \end{cases}$$

$$a_j = \begin{cases} 1 & \text{if AP } j \text{ is selected to transmit at 1.0 Mbps rate} \\ 0 & \text{otherwise.} \end{cases}$$

$$b_j = \begin{cases} 1 & \text{if AP } j \text{ is selected to transmit at 2.0 Mbps rate} \\ 0 & \text{otherwise.} \end{cases}$$

$$c_j = \begin{cases} 1 & \text{if AP } j \text{ is selected to transmit at 5.5 Mbps rate} \\ 0 & \text{otherwise.} \end{cases}$$

$$d_j = \begin{cases} 1 & \text{if AP } j \text{ is selected to transmit at 11.0 Mbps rate} \\ 0 & \text{otherwise.} \end{cases}$$

It is important to note that r_{ij} is a pre-computed value and therefore, not an optimization variable. Also τ is a predefined threshold value. The multicast association problem can be represented by the following integer programming problem, where the objective function is non-linear but all the constraints are linear.

$$\text{Maximize} \sum_{j=1}^{n} (r_j^{min} \times m_j)$$

subject to the following constraints:

$$\sum_{j \in S_{AP}} x_{ij} = 1 \ \forall i \in S_{STA} \tag{1}$$

$$\sum_{j \in S_{AP}} x_{ij} \ r_{ij} \geq \tau \ \forall i \in S_{STA} \tag{2}$$

$$\infty(1 - x_{ij}) + x_{ij} \ r_{ij} \geq r_j^{min} \ \forall i \in S_{STA}, \ j \in S_{AP} \tag{3}$$

$$m_j = \sum_{i \in S_{STA}} x_{ij} \ \forall j \in S_{AP} \tag{4}$$

$$r_j^{min} = 1 \ a_j + 2 \ b_j + 5.5 \ c_j + 11 \ d_j \ \forall j \in S_{AP} \tag{5}$$

$$a_j + b_j + c_j + d_j = 1 \ \forall j \in S_{AP}. \tag{6}$$

$$r_j^{min} \geq \tau \ \forall j \in S_{AP} \tag{7}$$

Here ∞ as used in Constraint (3) represents a big positive integer. Constraint (1) ensures that each STA should be associated with exactly one AP. Constraint (2) ensures that STA i can be associated with AP j only if $r_{ij} \geq \tau$. Constraint (3) ensures that r_j^{min} is set to the minimum rate among the rates obtained by all the STAs associated to AP j. Constraint (4) computes the number of STAs associated with AP j. Constraints (5) and (6) together ensure that the value of r_j^{min} belongs to the set of available rates $\mathscr{C} = \{1.0, 2.0, 5.5, 11.0\}$. Constraint (7) ensures that r_j^{min} must be greater than or equal to τ. The objective function represents the overall system throughput.

5 Motivational Examples and the Solution Approach

Our objective is to find an optimal association between the subscribed STAs and the available APs such that the overall system throughput is maximized. To demonstrate the impact of association on the overall system throughput, we consider the following examples.

Fig. 1. Motivational Example1: STA 2 gets the same data rate from AP 1 and AP 2.

Consider an network with 2 APs (AP 1 and AP 2) and 4 STAs (STA 1, STA 2, STA 3 and STA 4) as shown in Fig. 1. Here APs and STAs are shown by the filled circles and stars, respectively. The label associated with the edge (solid or dotted) between an STA and an AP indicates the physical rate (in Mbps) at which the STA can be associated with the AP. An association of an STA is termed as *fixed association* if the STA is under the coverage range of a single AP. It can be seen from Fig. 1 that STA 1, STA 3 and STA 4 have fixed associations with AP 1, AP 2 and AP 2 respectively. These fixed associations are shown as solid edges in Fig. 1. If an STA is under the coverage range of several APs, the STA can potentially be associated with any one among them. These potential associations are shown as dotted edges in Fig. 1. Note that though an STA may potentially be associated with many APs, but it must select *only one* AP from them for its association. It can be seen from Fig. 1 that STA 2 has two potential associations, one with AP 1 and the other with AP 2. So STA 2 must select either AP 1 or AP 2 for its association. It is to be noted that STA 2 gets the same data rate (2.0 Mbps) from both AP 1 and AP 2.

If STA 2 selects AP 1 to associate with, then the throughput provided by AP 1 will become (2.0×2) Mb as $r_1^{min} = 2.0$ Mbps and $m_1 = 2$. Similarly, the throughput provided by AP 2 will become (5.5×2) Mb. Hence the overall system throughput will become $(2.0 \times 2) + (5.5 \times 2) = 15.0$ Mb. But, if STA 2 selects AP 2 to associate with, then the overall system throughput will become $(5.5 \times 1) + (2.0 \times 3) = 11.5$ Mb. It is now clear that the first association of STA 2 provides more overall system throughput than the second one though STA 2 gets the same data rate from both the APs. Therefore, it is evident that the selection of an AP for association of an STA plays an important role for the overall system throughput even if the STA gets the same data rate from multiple APs.

We now consider a situation where an STA can potentially be associated with different APs at different rates. For this purpose we consider another example shown Fig. 2. In this situation, if STA 2 selects AP 1 to associate with, then the overall system throughput will become $(2.0 \times 2) + (1.0 \times 2) = 6.0$ Mb. But, if STA 2 selects AP 2 to associate with, then the overall system throughput will become $(2.0 \times 1) + (1.0 \times 3) = 5.0$ Mb. It is now clear that the association of STA 2 with AP 1 provides more overall system throughput than its association with AP 2 though STA 2 gets higher data rate from AP 2 than AP 1. It shows that the association based on RSSI, where an STA associates with an AP from which it gets the highest data rate, may not always provide good overall system throughput. In other words, maximizing r_j^{min} without taking care of m_j may not always produce the best result.

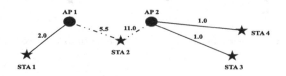

Fig. 2. Motivational Example2: STA 2 gets different data rates from AP 1 and AP 2.

We now consider an association policy based on *in-range STA number*, where an STA associates with an AP which has the maximum number of STAs in its coverage range. For this we consider the example shown in Fig. 1 again. From Fig. 1 it is evident that according to this association policy, STA 2 will select AP 2 for its association. As per our earlier discussion, association of STA 2 to AP 1 produces better overall system throughput than its association to AP 2. Hence the association based on *in-range STA number* may not always provide good overall system throughput. In other words, maximizing m_j without taking care of r_j^{min} may not always be the best option.

Motivated by the above observations, in our approach, we consider r_j^{min} and m_j simultaneously to maximize the overall system throughput instead of considering them independently. In the following section we now present our proposed approach formally.

6   The Proposed Greedy Algorithm

In this section, we present our proposed algorithmic solution to find an appropriate association between the multicast subscribed STAs and the available APs for providing multicast services to them. The objective is to maximize the overall system throughput while taking care of the user fairness.

Our approach works as follows. First we find the association of the STAs having the fixed associations. Then we calculate the overall system throughput considering the fixed associations only. Next we consider the association of the STAs having multiple potential associations. First we consider the set of STAs which get the highest data rate in $\mathscr{C}$ from at least one AP. The association of such an STA to a particular AP may pull up or pull down the current value of the overall system throughput. For an STA, we calculate the overall system throughput obtained from each such potential association and then choose the one which results in the highest pull ups or in the lowest pull downs. Next we consider the set of STAs which get the next lower data rate in $\mathscr{C}$ from at least one AP. The process is repeated until all STAs are covered. The detailed step by step description of the proposed greedy algorithm is stated below.

Input and Output: The proposed greedy algorithm takes the set of APs (S_{AP}), the set of STAs (S_{STA}), the set of available data rates $(\mathscr{C})$, the value of τ, the data rate matrix $R = (r_{ij})$ as inputs and returns the association between the STAs and the APs $(A = (a_{ij}))$ and the overall system throughput as outputs. In the resulted association matrix $A = (a_{ij})$, $a_{ij} = 1$ denotes that STA i is associated with AP j, and 0, otherwise.

Step 1: Initialization
Initially, no STA is being associated with and hence the values of r_j^{min}, m_j and σ_j are all set to zero for each AP $j \in S_{AP}$. This implies that initially the overall system throughput σ of the current multicast session is also zero. The association matrix is set to all zeros initially. Sort the available data rates in $\mathscr{C}$ in ascending order of their magnitudes and let $\mathscr{C}_1, \mathscr{C}_2, \cdots, \mathscr{C}_k$ be this sorted order where k is the cardinality of $\mathscr{C}$.

Step 2: Consideration of Fixed Associations

Step 2.1: Compute C_i for all STA i in S_{STA} where C_i is the set of APs from which STA i gets at least τ data rate. That is, compute $C_i = \{$AP j : $r_{ij} \geq \tau$ and $j \in S_{AP}\}$ for all $i \in S_{STA}$.

Remark 1. It is important to note that STA i must be associated with an AP which belongs to set C_i. If $|C_i| = 0$, STA i can not be associated to any AP of the network and thus will remain uncovered.

Step 2.2: Find $S_{USTA} = \{$STA i : $|C_i| = 0$ and $i \in S_{STA}\}$ where S_{USTA} is the set of uncovered STAs. Eliminate all such uncovered STAs from S_{STA} and update the set S_{RSTA} of remaining STAs as $S_{RSTA} = S_{STA} \setminus S_{USTA}$.

Remark 2. Association of STA i is termed as *fixed association* if $|C_i| = 1$. In such case, since $|C_i| = 1$, STA i must be associated with the only AP in C_i.

Step 2.3: Find $S_{FSTA} = \{$STA i : $|C_i| = 1$ and $i \in S_{STA}\}$ where S_{FSTA} is the set of STAs having fixed associations. Find $S_{EAP} = \cup_{i \in S_{FSTA}} C_i$ where S_{EAP} is the set of APs each of which covers at least one STA having the fixed association with it.

Step 2.4: Associate the STAs in S_{FSTA} to their respective APs in S_{EAP}. After making all the fixed associations, update the association matrix A and compute the values of r_j^{min}, m_j and σ_j for each AP j in S_{EAP}. Also compute the overall system throughput $\sigma = \sum_{j \in S_{AP}} \sigma_j$. Remove the STAs in S_{FSTA} from the network and update $S_{RSTA} = S_{RSTA} \setminus S_{FSTA}$. If $S_{RSTA} = \emptyset$ then the algorithm is terminated, otherwise, go to the next step for considering the remaining STAs in S_{RSTA} having multiple potential associations.

Remark 3. The STAs in S_{RSTA} having multiple potential associations are associated with the APs with a view to maximizing the overall system throughput. However, to provide the fairness towards the individual throughput obtained by the STAs, we first associate the STAs which get the highest data rate $\mathscr{C}_k$ from at least one AP. Then we consider the STAs which get the next lower data rate $\mathscr{C}_{k-1}$ and so on until $S_{RSTA} = \emptyset$. In this way we maximize the overall system throughput while respecting the fairness of the individual throughput of the STAs.

Step 3: Consideration of Multiple Potential Associations

Step 3.1: Compute $D_j^k = \{$STA i : $r_{ij} = \mathscr{C}_k$ and $i \in S_{RSTA}\}$ for all $j \in S_{AP}$ where D_j^k is the set of STAs in S_{RSTA} which get the data rate $\mathscr{C}_k$ from AP j. Compute $S_{PSTA} = \cup_{j \in S_{AP}} D_j^k$ where S_{PSTA} is the set of STAs which get the data rate $\mathscr{C}_k$ from at least one AP in S_{AP}.

Step 3.2: In this step, we find the association of STA i in S_{PSTA} to an appropriate AP j in C_i which results in the highest pull ups or in the lowest pull downs of the overall system throughput. Let σ_j be the previous throughput provided by AP j before considering the association of STA i to it. Compute σ'_j, the throughput that can be provided by AP j, if STA i is associated with AP j. The association of STA i to AP j may pull up or pull down the previous throughput σ_j provided by AP j. Compute $\sigma_j^{cost} = (\sigma'_j - \sigma_j)$. If $\sigma_j^{cost} \geq 0$, association of STA i to AP j will pull up the previous throughput σ_j and hence it will pull up the overall system throughput σ as well. Else if $\sigma_j^{cost} < 0$, such association will pull down σ. After computing σ_j^{cost} for all $j \in C_i$, associate STA i to that AP which provides highest pull ups or lowest pull downs. That is, associate STA i to AP j' if $\sigma_{j'}^{cost} = \max_j\{\sigma_j^{cost} : j \in C_i\}$. If multiple such APs are found then break the ties based on the higher value of r_{ij}, lower value of m_j and finally the lower value of AP index. After associating STA i to AP j', update the association matrix A and compute the values of $r_{j'}^{min}$, $m_{j'}$, $\sigma_{j'}$ and σ. Remove STA i from both S_{PSTA} and S_{RSTA} and update $S_{PSTA} = S_{PSTA} \setminus \{\text{STA } i\}$ and $S_{RSTA} = S_{RSTA} \setminus \{\text{STA } i\}$ accordingly. Repeat this step until $S_{PSTA} = \emptyset$.

Step 4: Consideration of Different Data Rates
If $S_{RSTA} = \emptyset$ then the algorithm is terminated, otherwise, set $k = k - 1$ and repeat Step 3 until $S_{RSTA} = \emptyset$.

Time Complexity: The time complexity of the proposed algorithm is $\mathcal{O}(nm^2k)$ where n, m and k are the cardinalities of S_{AP}, S_{STA} and $\mathscr{C}$ respectively.

Remark 4. The proposed algorithm is a centralized algorithm where all the input data needs to be known before execution of the algorithm. Each AP in the network monitors the spectrum and measures the channel condition at a regular interval. This allows the AP to find the number of STAs which are present within its coverage range and also to estimate the data rates that they may get from it. Each AP will send this information to the network controller through the wired backbone network. After receiving this information from all the available APs, the network controller will be able to execute the proposed greedy algorithm.

7 Performance Evaluation

In this section, we evaluate the performance of our proposed greedy algorithm and compare the results with other well-known metrics.

7.1 Simulation Set-Up

We have considered an infrastructure based IEEE 802.11b WLAN where a number of APs and a number of STAs are uniformly placed in an $1000 \times 1000\,\text{m}^2$ area. We vary the number of APs from 30 to 300 with a step of 10 and the number of STAs is varying from 50 to 1000 with a step of 50. The MAP is placed at the position $(0, 0)$ which is the lower left most corner of the considered area. The coverage and interference range of each AP are set to 150 and 240 m respectively. We assume a simple wireless channel model for our simulation where the data rate obtained by a subscribed STA depends on the distance of it from the serving AP [5,6]. The STAs which are within 50 ms from an AP will get 11.0 Mbps, 5.5 Mbps between 50 and 80 m, 2.0 Mbps between 80 and 120 m and when the distance is between 120 and 150 m the data rate is 1.0 Mbps [5,6]. These values are commons with those provided by IEEE 802.11 vendors [27]. An STA will get 0.0 data rate from an AP if it is located beyond the distance of 150 m from it. An STA is not considered to be part of the network if no AP is there within 150 m from it. We consider all the STAs as subscribed STAs for the multicast session under consideration.

7.2   Simulation Results

In this section, we have considered some well-known metrics and a metric namely *normalized cost* [15] to compare the performance of our proposed greedy algorithm. In RSSI metric [4,5,7,8,14–16], an STA is associated with an AP from which it gets the maximum data rate. The *minimum hop-distance* metric tells that an STA will be associated with that AP which has the minimum hop-distance to reach the MAP [7,8,15,16]. In *in-range STA number* metric [7,8,15,16], an STA will be associated with that AP which has the maximum number of STAs in its coverage range. Apart from these three well-known metrics, we also have considered a metric namely *normalized cost* used in [7,8,15]. The normalized cost of an AP is defined as the ratio $\frac{H}{N}$, where H is the minimum hop-distance to reach the MAP from it and N is the number of STAs in its range. In normalized cost metric, an STA will be associated with that AP which has the minimum normalized cost value. Apart from these metrics, we also have compared the performance of our algorithm with the unirate multicasting approach where all the APs transmit their respective multicast data packets at the same basic data rate $r^{min} = \min_j\{r_j^{min} : j \in S_{AP}\}$. The theoretical maximum value of the overall system throughput can be expressed as $\sigma^{max} = \sum_{i \in S_{STA}} t_i$, where $t_i = \max_j\{r_{ij} : j \in S_{AP}\}$ is the largest possible rate STA i can be associated with. It is to be noted that σ^{max} represents a naive upper bound of the overall system throughput which is independent of any association strategy. In fact σ^{max} may not be achievable by any association policy in practice.

In fact there may not exist any association which achieves this σ^{max}.

Table 1 shows the comparison of the results obtained by the proposed algorithm against the said metrics. In this simulation, we have placed 50 APs and 210 STAs uniformly over the considered area. For each result, we have considered 100 different placements of the APs and the STAs and report their average value. Figure 3 shows one such instance where circles denote the positions of the APs and asterisks denote the positions of the STAs. The MAP is placed at the position $(0,0)$ which is at lower left most corner of the considered area. It can be seen from Table 1 that proposed algorithm provides 27%, 196%, 206%, 239% and 319% more overall system throughput than RSSI, minimum hop-distance, in-range STA number, normalized cost and unirate multicasting respectively, when $\tau = 1.0$ Mbps. When $\tau = 2.0$ Mbps, our proposed algorithm gives 37%, 83%,

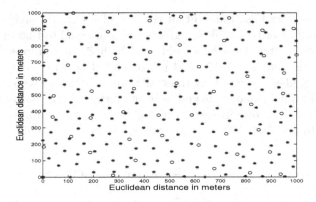

Fig. 3. A possible positions of the considered 50 APs and 210 STAs.

75%, 84% and 128% more system throughput than these metrics, respectively. Note that for the value of $\tau = 1.0$ Mbps, all STAs are served, but for the values of $\tau = 2.0, 5.5$ and 11.0 Mbps, 99.61%, 79.42% and 35.09% of STAs are served by all these metrics and the proposed algorithm. However, the proposed algorithm provides 33%, 33%, 31%, 31% and 41% more overall system throughput than these metrics, when $\tau = 5.5$ Mbps. And when $\tau = 11.0$ Mbps, all metrics and proposed algorithm give the same amount of overall system throughput.

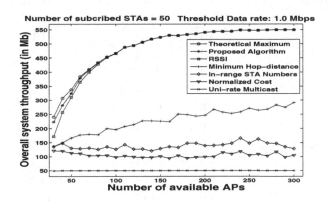

Fig. 4. Number of available APs vs overall system throughput when number of STAs is 50 and $\tau = 1$ Mbps.

Figures 4 and 5 show how the overall system throughput is varying with the number of available APs for a fixed number of subscribed STAs. In Figs. 4 and 5, we have considered 50 and 300 STAs respectively, to represent different traffic load distributions. The number of APs is varying from 30 to 300 with a step of 10 to represent different network densities. The coverage range, interference range and the value of τ are set at 150 m, 240 m and 1.0 Mbps respectively.

It is evident from Figs. 4 and 5 that the overall system throughput obtained by the proposed algorithm is always greater than or equal to that of other metrics. The overall system throughput obtained by RSSI and proposed algorithm increase with the number of APs and get saturated at some point. This saturation point signifies the fact that those APs are sufficient to serve each STA at the maximum available data rate and hence no further improvement is observed after that point. The overall system throughput at the saturation points are 550 (50 × 11.0) and 3300 (300 × 11.0) Mb as shown in Figs. 4 and 5 respectively.

Table 1. Performance comparison of the proposed algorithm with different metrics.

Name of metric	Value of τ (in Mbps)	Overall system throughput (in Mb)	% of throughput improvement
Theoretical maximum (σ^{max})	1.0	1408.35	
	2.0	1407.55	
	5.5	1322.75	
	11.0	810.7	
Proposed algorithm	1.0	880.3	
	2.0	955.5	
	5.5	1294.7	
	11.0	810.7	
Unirate multicast	1.0	210	319.19
	2.0	418.4	128.36
	5.5	917.4	41.12
	11.0	810.7	0
$RSSI$	1.0	691.75	27.25
	2.0	696.85	37.11
	5.5	972.4	33.14
	11.0	810.7	0
Minimum Hop-distance	1.0	296.55	196.84
	2.0	520.15	83.69
	5.5	971.85	33.22
	11.0	810.7	0
In-range STA number	1.0	286.85	206.88
	2.0	544.3	75.54
	5.5	987.25	31.14
	11.0	810.7	0
Normalized cost	1.0	259.35	239.42
	2.0	518.55	84.26
	5.5	984.5	31.50
	11.0	810.7	0

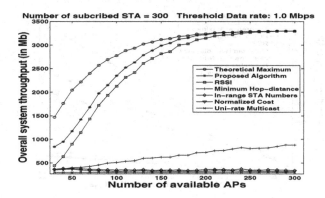

Fig. 5. Number of available APs vs overall system throughput when number of STAs is 300 and $\tau = 1$ Mbps.

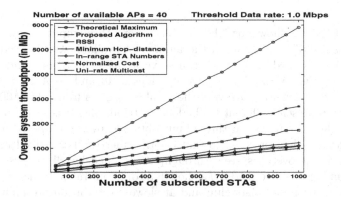

Fig. 6. Number of subscribed STAs vs overall system throughput when number of APs is 40 and $\tau = 1$ Mbps.

It is seen from Figs. 4 and 5 that both proposed algorithm and RSSI touch the theoretical maximum value. However, the proposed algorithm touches the theoretical maximum value earlier than the RSSI. It can be observed that the overall system throughput provided by minimum hop-distance increases as the number of APs increases but the rate at which it increases is very low. The overall system throughput more or less remains constant in both in-range STA number and normalized cost. However, the magnitude of overall system throughput in in-range STA number is slightly higher than that of normalized cost. It is seen from Figs. 4 and 5 that the value of overall system throughput obtained by unirate multicasting remains constant at 50 (50 × 1.0) and 300 (300 × 1.0) Mb respectively. This happens because in unirate multicasing, multicast packets are transmitted at the same basic data rate (1.0 Mbps) to all the subscribed STAs. So we can infer that the overall system throughput not only depends on the relative positions of APs and STAs but also on the association strategy.

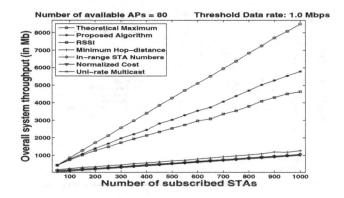

Fig. 7. Number of subscribed STAs vs overall system throughput when number of APs is 80 and $\tau = 1$ Mbps.

Figures 6 and 7 show how the overall system throughput varies with the number of subscribed STAs when the number of available APs is fixed. In this case we have assumed that an AP can serve at most 32 STAs simultaneously [15,16,29]. In Figs. 6 and 7, we have considered 40 and 80 APs to represent different network densities. We vary the number of STAs from 50 to 1000 with a step of 50 to represent different traffic load distributions. It is seen from Figs. 6 and 7 that the overall system throughput obtained by the proposed algorithm is always greater than or equal to that of other metrics. For a fixed number of available APs, the overall system throughput obtained by all metrics, increase with the number of subscribed STAs though at different rates. It increases at the fastest rate in proposed algorithm and at slowest rate in unirate multicasting. The difference between the overall system throughput obtained by these metrics and the theoretical maximum increases with the number of subscribed STAs. This difference decreases with the increase in the number of available APs as evident from Figs. 6 and 7. This signifies the fact that we have to place a sufficiently

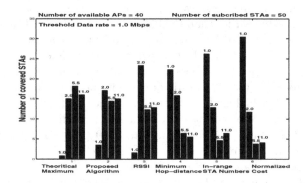

Fig. 8. Number of STAs (50) covered at different rates for different approaches when number of available APs is 40 and $\tau = 1$ Mbps.

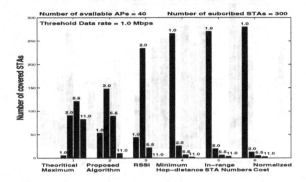

Fig. 9. Number of STAs (300) covered at different rates for different approaches when number of available APs is 40 and $\tau = 1$ Mbps.

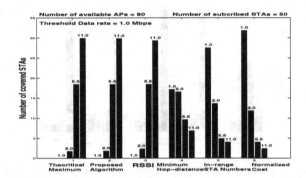

Fig. 10. Number of STAs (50) covered at different rates for different approaches when number of available APs is 80 and $\tau = 1$ Mbps.

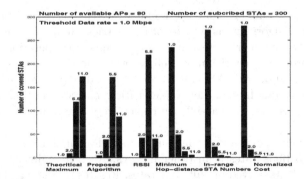

Fig. 11. Number of STAs (300) covered at different rates for different approaches when number of available APs is 80 and $\tau = 1$ Mbps.

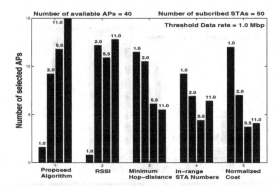

Fig. 12. Number of selected APs (40) at different rates for different approaches when number of STAs is 50 and $\tau = 1$ Mbps.

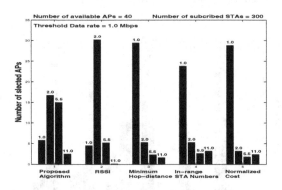

Fig. 13. Number of selected APs (40) at different rates for different approaches when number of STAs is 300 and $\tau = 1$ Mbps.

large number of APs to achieve throughput equal to the theoretical maximum, i.e., to reduce this difference to zero.

Figures 8, 9, 10 and 11, show the number of STAs served at different rates by our proposed algorithm as well as other metrics. Similarly Figs. 12, 13, 14 and 15, show the number of APs are being selected for operation at different rates by our proposed algorithm as well as other metrics. In these figures, we have considered 50 or 300 STAs and 40 or 80 APs to represent different traffic load distributions and different network densities. The coverage and interference range of each AP is set to 150 and 240 m respectively. The value of τ is set at 1.0 Mbps.

It is seen from Figs. 8 and 9 that our proposed algorithm serves more STAs at 11.0 and 5.5 Mbps rates than other metrics. Also, it maintains a balanced distribution of STAs served at different rates. Figures 10 and 11 also show the similar trend but the magnitude of the number of STAs served at different rates are different. From Figs. 8, 9, 10 and 11, we can conclude that our proposed algorithm maintains a good amount of fairness between the individual throughput

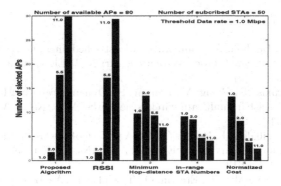

Fig. 14. Number of selected APs (80) at different rates for different approaches when number of STAs is 50 and $\tau = 1$ Mbps.

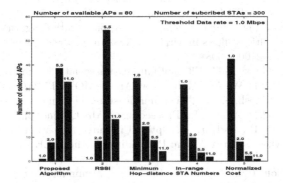

Fig. 15. Number of selected APs (80) at different rates for different approaches when number of STAs is 300 and $\tau = 1$ Mbps.

obtained by the STAs. From Figs. 12, 13, 14 and 15, it can be observed that our proposed algorithm selects more number of APs to operate at higher data rates, than other metrics. It also maintains a balanced distribution of selected APs operated at different data rates.

8 Conclusion

An efficient greedy algorithm to find an optimal association for multirate multicasting in WLAN is developed which maximizes the overall system throughput while taking care of the user fairness. We have evaluated and compared the performance of the proposed algorithm with other well-known metrics. The obtained results show that the proposed algorithm significantly improves the overall system throughput in comparison to these metrics. Our future work is to implement the proposed algorithm in a distributed as well as online setup.

References

1. Afolabi, R.O., Dadlani, A., Kim, K.: Multicast scheduling and resource allocation algorithms for OFDMA-based systems: a survey. IEEE Commun. Surv. Tutor. **15**(1), 240–254 (2013)
2. Alay, O., Korakis, T., Wang, Y., Panwar, S.: Dynamic rate and FEC adaptation for video multicast in multirate wireless networks. Mob. Netw. Appl. (MONET) **15**(3), 425–434 (2010)
3. Santos, M.A., Villalon, J., Barbosa, L.O.: A novel QoE-aware multicast mechanism for video communications over IEEE 802.11 WLANs. IEEE J. Sel. Areas Commun. **30**(7), 1205–1214 (2012)
4. Athanasiou, G., Korakis, T., Ercetin, O., Tassiulas, L.: A cross-layer framework for association control in wireless mesh networks. IEEE Trans. Mob. Comput. **8**(1), 65–80 (2009)
5. Bejerano, Y., Han, S.J., Li, L.: Fairness and load balancing in wireless LANs using association control. IEEE/ACM Trans. Netw. **15**(3), 560–573 (2007)
6. Bejerano, Y., Lee, D., Sinha, P., Zhang, L.: Approximation algorithms for scheduling real-time multicast flows in wireless LANs. In: Proceedings of the IEEE INFOCOM, pp. 2092–2100 (2008)
7. Bhaumick, D., Ghosh, S.C.: Efficient multicast association to improve the throughput in IEEE 802.11 WLAN. Mob. Netw. Appl. (MONET) **21**(3), 436–452 (2016)
8. Bhaumick, D., Ghosh, S.C.: Efficient multicast association to improve the throughput in IEEE 802.11 WLAN. In: Proceedings of the QShine, pp. 83–89 (2014)
9. Bui, L., Srikant, R., Stolyar, A.: Optimal resource allocation for multicast flows in multihop wireless networks. In: Proceedings of the IEEE CDC, pp. 1134–1139 (2007)
10. Qadir, J., Chou, C.T., Misra, A., Lim, J.G.: Minimum latency broadcasting in multiradio, multichannel, multirate wireless meshes. IEEE Trans. Mob. Comput. **8**(11), 1510–1523 (2009)
11. Deb, S., Srikant, R.: Congestion control for fair resource allocation in networks with multicast flows. IEEE/ACM Trans. Netw. **12**(2), 274–285 (2004)
12. Kar, K., Sarkar, S., Tassiulas, L.: Optimization based rate control for multirate multicast sessions. In: Proceedings of the IEEE INFOCOM, pp. 123–132 (2001)
13. Kar, K., Sarkar, S., Tassiulas, L.: A scalable low-overhead rate control algorithm for multirate multicast sessions. IEEE J. Sel. Areas Commun. **20**(8), 1541–1557 (2002)
14. Kumar, A., Kumar, V.: Optimal association of stations and APs in IEEE 802.11 WLAN. In: Proceedings of the NCC (2005)
15. Lee, D., Chandrasekaran, G., Sinha, P.: Optimizing broadcast load in mesh networks using dual-association. In: Proceedings of the IEEE Workshop on Wireless Mesh Networks (2005)
16. Lee, D., Chandrasekaran, G., Sridharan, M., Sinha, P.: Association management for data dissemination over wireless mesh networks. Comput. Netw. **51**(15), 4338–4355 (2007)
17. Li, C., Xiong, H., Zou, J., Wu, D.O.: Dynamic rate allocation and opportunistic routing for scalable video multirate multicast over time-varying wireless networks. In: Proceedings of the IEEE INFOCOM Workshop, pp. 275–280 (2014)
18. Mills, D.L.: On the accuracy and stability of clocks synchronized by the network time protocol in the internet system. ACM SIGCOMM Comput. Commun. Rev. **20**(1), 65–75 (1990)

19. Paschos, G.S., Li, C., Modiano, E., Choumas, K., Korakis, T.: Multirate multicast: optimal algorithms and implementation. In: Proceedings of the IEEE INFOCOM, pp. 343–351 (2014)
20. Sarkar, S., Tassiulas, L.: A framework for routing and congestion control for multicast information flows. IEEE Trans. Inf. Theory **48**(10), 2690–2708 (2002)
21. Sarkar, S., Tassiulas, L.: Fair allocation of utilities in multirate multicast networks: a framework for unifying diverse fairness objectives. IEEE Trans. Autom. Control. **47**(6), 931–944 (2002)
22. Sarkar, S., Tassiulas, L.: Fair distributed congestion control in multirate multicast networks. IEEE Trans. Netw. **13**(1), 121–133 (2005)
23. Sarkar, S., Tassiulas, L.: Fair allocation of discrete bandwidth layers in multicast networks. In: Proceedings of the IEEE INFOCOM, pp. 1491–1500 (2000)
24. Suh, C., Mo, J.: Resource allocation for multicast services in multicarrier wireless communications. IEEE Trans. Wirel. Commun. **7**(1), 27–31 (2008)
25. Zhao, X., Guo, J., Chou, C.T., Misra, A., Jha, S.: A high-throughput routing metric for reliable multicast in multirate wireless mesh networks. In: Proceedings of the IEEE INFOCOM (2011)
26. IEEE Std 802.11-2012. http://standards.ieee.org/getieee802/download/802.11-2012.pdf
27. Enterprise mobility 7.3 design guide, September 2013. http://www.cisco.com/c/en/us/td/docs/solutions/Enterprise/Mobility/emob73dg/emob73.pdf
28. Data Sheet for Cisco Aironet 1200 Series, Cisco Systems Inc. (2004)
29. http://www.cisco.com/c/en/us/products/collateral/wireless/aironet-1250-series/design_guide_c07-693245.pdf
30. ORINOCO AP-600 Data Sheet, ProximWireless Networks (2004)
31. XG-705S specification. http://www.zcomax.co.uk/doc/XG-705S%20Draft%20Product%20Specification_C0_060517.pdf
32. AG-623C IEEE 802.11 a/b/g miniPCI specification. http://www.zcomax.co.uk/doc/AG-623C.pdf

An ns-3 MPTCP Implementation

Kashif Nadeem and Tariq M. Jadoon[(✉)]

Electrical Engineering Department,
Syed Babar Ali School of Science and Engineering,
Lahore University of Management Sciences, Lahore, Pakistan
kshfnadeem@gmail.com, jadoon@lums.edu.pk

Abstract. Multipath TCP (MPTCP) achieves greater throughput by sending packets from a single byte stream across multiple interfaces and thus, potentially exploits multiple available network paths. This allows end hosts to aggregate bandwidth and network resources. Network simulators such as ns-3 [1] provide researchers with a convenient tool to evaluate protocols and architectures and their importance can not be overemphasized. There are currently 3 existing implementations of MPTCP in ns-3. We evaluate these implementations and find that they lack several key features and are therefore, inadequate for furthering research. We implement MPTCP in ns-3-dev (Developer's version) and introduce multiple path managers namely *default*, *ndiffports* and *fullmesh* creating an MPTCP patch for ns-3 [2]. The simulation results show improvements in throughput and Flow Completion Times (FCTs) in comparison with previous work. Our implementation [3] is compatible with the current version (ns-3.29).

Keywords: MPTCP · ns-3 · Computer networks · Simulator

1 Introduction

Applications such as Facebook, Google, etc., require low latency and place bounds on the response time of a query initiated by a user. Excessive delays in response time of queries impact revenue and user experience. Transmission Control Protocol (TCP) [4] fails to provide high throughput to large flows and complete latency sensitive flows within time bounds. As TCP only uses a single interface of an end host even if other interfaces available, it thus, under utilizes network resources. In recent years, a number of transport layer protocols have been proposed to improve throughput such as, DCTCP [5], D2TCP [6], PIAS [7], etc. However, these protocols also use a single interface and do not exploit multiple interfaces even if available.

Modern network devices such as computers, smart phones and tablets typically have more than one network interface and can thus, be multihomed. Smart phones have wi-fi and 3G/4G interfaces whilst, laptops have Ethernet and wi-fi

T. Q. Duong et al. (Eds.): Qshine 2018, LNICST 272, pp. 48–60, 2019.
https://doi.org/10.1007/978-3-030-14413-5_4

interfaces. Previous studies have shown that simultaneously using multiple interfaces can achieve higher throughput and can complete large flows in a shorter time [8–11]. Multipath TCP uses the available interfaces of a network device to send application data to the destination through multiple network paths. MPTCP splits the application data stream amongst subflows whereby, each subflow follows a path based on a 5-tuple: source IP, destination IP, source port, destination port and layer-4 protocol. These sub-flows can be routed over different paths by using routing protocols such as Equal-Cost Multi-Path (ECMP) [12]. The subflows can follow the same network path or different paths depending upon the availability of paths between the two hosts. MPTCP aggregates the bandwidth of available links to the hosts by applying the concept of Resource Pooling i.e., the available resources or links appear as a single logical resource or link to the host [8]. Multipath TCP has been implemented in several operating systems such as Linux, Apple ios7, Mac and FreeBSD. Furthermore, MPTCP can be deployed in data centers [13] and better exploits data center topology, effectively utilizes available bandwidth and provides improved throughput.

The balance of the paper is organised as follows. In Sect. 2 we describe the architecture and design of MPTCP from RFCs. In Sect. 3, we discuss the available implementations of MPTCP in different ns-3 versions. Furthermore, we discuss compatibility issues and missing features of these implementations. In Sect. 4, we briefly discuss why there is a need to implement MPTCP in the latest ns-3 tree and provide an overview about changes in TCP classes and discuss our implementation procedure. In Sect. 5, we describe simulation experiments to compare our implementation with Coudron et al. [14]. We finish with conclusions and directions for future work.

2 MPTCP Details from RFCs

This section provides a brief overview of MPTCP architecture, design and compatibility issues.

2.1 MPTCP Architecture and Path Managers

RFC 6182 [15] provides an architectural overview of MPTCP and discusses compatibility challenges with the existing network stack and middle boxes. MPTCP explores possible paths by using all available interfaces (IPs) at source and destination hosts.

In Fig. 1, Host A and B have two network interfaces each with unique IP addresses. MPTCP can thus, exploit four possible paths: A1-B1, A1-B2, A2-B1 and A2-B2 and moreover, can establish subflows on each path. These paths are not necessarily disjoint therefore, subflows may traverse the same link or the same path. The number of available paths depends upon the underlying physical network between the end hosts. MPTCP conveys protocol specific information in the header through the TCP options field. Middle boxes such as Network Address Translators (NATs), Performance Enhancing Proxies (PEPs), Intrusion

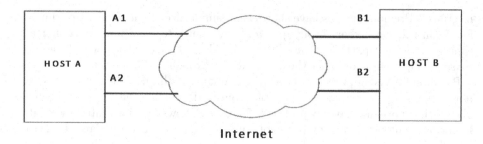

Fig. 1. Simple multipath scenario

Detection Systems (IDs) and Firewalls can potentially rewrite the TCP options or may drop a connection upon seeing unknown TCP options. RFC 3234 [16] mentions that a protocol designed without considering middle boxes can fail in the presence of middle boxes and suggests architectural design goals for new protocols and middle boxes. MPTCP design enforces the consideration of middle boxes so that subflows appear as legacy TCP flows to middle boxes. RFC 6897 [17] describes compatibility issues of MPTCP with existing applications and provides an application interface for MPTCP to work with legacy applications. This new design is based on the "Transport next-generation" (Tng) model [18]. The Tng model splits the transport layer into a "Semantic" layer which supports and implements functionality for the application layer and a "Flow+Endpoint" layer which manages the network-oriented part of the transport layer. The implementation of a network-oriented part in the transport layer enables the end hosts to interact with middle boxes as if they are working with legacy TCP. Figure 2 shows the Tng decomposed model of the internet and MPTCP protocol stack.

Application		Application	
Semantic		MPTCP	
Flow+ End-Point	Flow+ End-Point	Subflow (TCP)	Subflow (TCP)
Network	Network	IP	IP

Fig. 2. Tng model (left) vs MPTCP model (right)

MPTCP implements several functions such as path management, packet scheduling, subflow interface and congestion control to create and manage subflows among multihomed hosts. Path Management is a core function of MPTCP which upon initial setup of the MPTCP connection creates subflows between the two end hosts. The Linux Kernel Implementation [19] provides four path managers. Users can select any one of them at compile time:

- **Default:** In default mode, the path management mechanism doesn't create new subflows. Hosts neither advertise IP addresses nor create new subflows however, passive creation of subflows is supported.
- **Fullmesh:** With this Path manager, multihomed hosts advertise addresses to peers and create a complete mesh of new subflows across all possible pairs of IP addresses. Considering the scenario depicted in Fig. 1; a fullmesh path manager will create four subflows between IP pairs: A1-B1, A1-B2, A2-B1 and A2-B2. Thus, the number of subflows is limited by number of the IP pairs.
- **ndiffPorts:** This path manager initiates subflows between the same IP pair using different source and destination ports. It can hence create any number of subflows between a pair of IP addresses such as A1-B1 shown in Fig. 1. The number of subflows created is controlled through a parameter.
- **Binder:** This path manager [20] uses Loose Source and Record Routing (LSRR) without modification of the end-user devices. Binder provides a list of available gateways to MPTCP subflows and ensures that subflows visit these gateways and explore all available paths in the network. The packets of subflows are distributed over the network using relays and proxies to explore available network paths.

2.2 MPTCP Design

RFC 6824 [21] provides a detailed description of MPTCP connection establishment, subflow initiation and MPTCP options used to carry information across the internet. Internet Assigned Numbers Authority (IANA) added a new TCP option for MPTCP with the symbolic name "Kind" with a 4-bit subtype field providing "MPTCP Option Subtypes". MPTCP option subtypes include MP_CAPABLE, MP_JOIN, ADD_ADDR, MP_FAIL, DSS, etc. Here we briefly discuss connection setup, subflow initiation and data sequence mapping.

- **MPTCP connection setup:** End hosts use the traditional 3-way TCP handshake mechanism SYN, SYN/ACK, ACK but each packet contains an MP_CAPABLE option as shown in Fig. 3. Moreover, the packets include a sender's key and a receiver's key that are used in future for the creation of new subflows. This MPTCP handshake ensures that the receiver and middle boxes are MPTCP capable. If any received packet doesn't contain the MP_CAPABLE option then it means that either the receiver or middle boxes are not MPTCP capable and the connection falls back to regular TCP.
- **Creating new subflows:** After the establishment of an MPTCP connection, hosts can create subflows when the first DATA_ACK is received through the Data Sequence Signaling (DSS) option. The connection initiation messages SYN, SYN/ACK and ACK include MP_JOIN. The sender sends a 32-bit token generated by SHA-1 from the receiver's key with the SYN packet to associate subflows with the MPTCP connection. Senders and receivers exchange

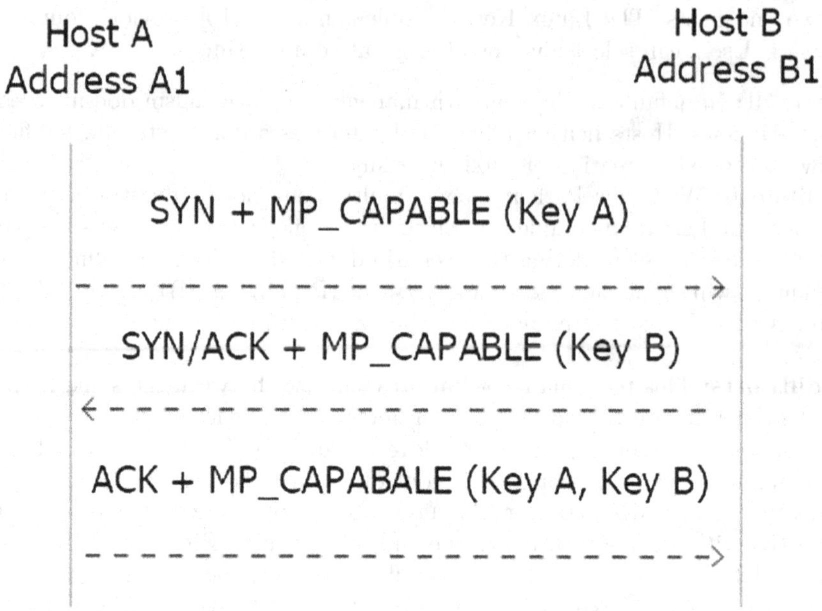

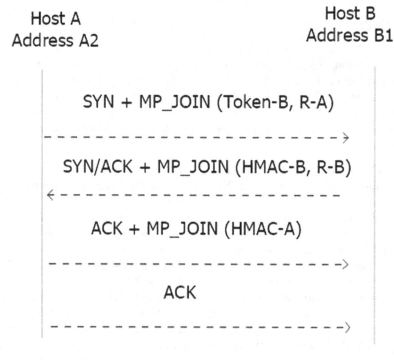

Host A
Address A1

Host B
Address B1

SYN + MP_CAPABLE (Key A)

SYN/ACK + MP_CAPABLE (Key B)

ACK + MP_CAPABALE (Key A, Key B)

Fig. 3. MPTCP 3-way handshake

Host A
Address A2

Host B
Address B1

SYN + MP_JOIN (Token-B, R-A)

SYN/ACK + MP_JOIN (HMAC-B, R-B)

ACK + MP_JOIN (HMAC-A)

ACK

Fig. 4. Subflow initiation from multi addressed Host A

a nonce and Hashed Message Authentication Code (HMAC) for connection authentication. Any packet received without a MP_JOIN will result in falling back to traditional TCP. The subflow initiation process between host A with address A2 and host B with address B1 is shown in Fig. 4.

3 Existing MPTCP Implementations in ns-3 and Their Shortcomings

This section briefly describes the available ns-3 implementations of MPTCP and discusses their shortcomings.

3.1 ns-3 Implementations of MPTCP

At present, three implementations of MPTCP are available in ns-3. MPTCP was first implemented in ns-3 by Chihani et al. [22] in ns-3.6. The TCP stack was rewritten in ns-3.8 which makes this implementation incompatible with later ns-3 versions and is now obsolete. In this implementation, the sender and receiver do not exchange keys to associate new subflows with the MPTCP connection. They also didn't follow an authentication process for subflows through nonces and HMACs as described in [21].

The second implementation of MPTCP is in ns-3.19 by Kheirkhah et al. [23] and follows the Linux kernel implementation of MPTCP [24]. The authors create a MpTcpSocketBase class; which is a subclass of the TcpSocketBase class. Upon a successful MPTCP connection a MPTCP socket is created that provides an interface between the application and TCP flows. The MpTcpSocketBase object controls path management, scheduling, packet reordering and congestion control algorithms, etc. Another class MpTcpSubflow defines TCP subflows that communicate with the network layer. These two MPTCP implementations [22, 23] lack several features such as backward compatibility with the TCP stack and support a subset of MPTCP options as listed in Table 1 in [14]. The ns-3.19 implementation doesn't implement HMAC based authentication for subflows as shown in Fig. 4 and also lacks support for TCP timestamps, window scale options, congestion control algorithms such as Scalable, H-TCP, TCP Vegas, etc. Moreover, ns-3.19 lacks many new TCP features making it less appealing to the research community. In later ns-3 versions, TCP classes have changed substantially which makes this implementation incompatible with newer ns-3 versions.

The third implementation of MPTCP is in ns-3.23 by Coudron et al. [14]. This implementation covers several deficiencies of the previous implementations as shown in Table 1 in [14]. The MPTCP connection starts with a TCP socket and the client sends a SYN + MP_CAPABLE option with its key. The server receives SYN packet and then upgrades to a MPTCP socket if it is MPTCP capable and replies with a SYN/ACK + MP_CAPABLE option along with its server key. The client upon receiving SYN/ACK upgrades to a MPTCP socket and replies with an ACK completing the 3-way MPTCP handshake. This MPTCP socket provides an interface between the application and TCP subflows for packet scheduling, reordering and retransmission, etc. Round robin and fastest RTT schedulers are implemented and divide the application byte stream into segments for transmission on established subflows.

3.2 Problems with the ns-3.23 [14] Implementation

Coudron et al. [14] MPTCP implementation faces two kinds of problems:

Missing and Incomplete Features. Although, this implementation covers several deficiencies of [22,23] as shown in Table 1 in [14] yet it still lacks several basic MPTCP components shown in Table 1. This implementation doesn't implement *path management* discussed in Sect. 2.1 which is a core MPTCP function. Without path managers MPTCP behaves just like single-path TCP and doesn't create subflows even if hosts are multi-homed and multi-addressed. Coudron et al. [14] implementation creates a master subflow to transmit an application byte stream but subsequently does not create further subflows. Although there are three classes for MPTCP congestion control i.e., MpTcpCongestionCoupled, MpTcpCCOlia and MpTcpCCUncoupled, they are incomplete and not functional. Furthermore, this implementation also lacks the MP_FAIL option, infinite mapping, checksums and MP_PRIO.

Compatibility Issues. ns-3 is a continuously evolving project wherein each new version potentially introduces new models and classes as well as modifies existing models and classes. ns-3.25 refactored TCP removing, modifying and appending some TCP classes, functions and variables. Furthermore, ns-3.25 introduced new congestion control classes as well as Active Queue Management (AQM), policing and packet filtering. Similarly, ns-3.26 introduced new congestion *control classes* such as TCP Vegas, Veno, H-TCP and Illinois, etc. for legacy TCP. Several new queueing models such as Linux-like pfifo_fast, FQ_CoDel, Byte Queue Limits, Adaptive RED have been added in the Traffic Control Module. Moreover, ns-3.26 implements Fast retransmit and Fast recovery as described in RFC 5681 [25]. ns-3.27 incorporates SACK and LEDBAT in the TCP model amongst other models. ns-3.28 added IPV6 support for LTE, TCP

Table 1. A comparison of features in various MPTCP implementations

Features	Kheirkha et al.	Coudron et al.	ns-3-dev MPTCP
Path managers	Default, fullmesh, ndiffPorts	None	Default, ndiffPorts, fullmesh
Subflow creation	Yes	No	Yes
Congestion control	Yes	No	No
Compatibility[a]	No	No	Yes

[a]Compatibility with current ns-3 stack

pacing, FIFO and TBF queue, etc. These changes in TCP classes make Coudron et al. [14] incompatible with the current ns-3 stack. Thus, older ns-3 versions lack many networking components and features making them less attractive for present research work.

4 Our MPTCP Implementation in ns-3-dev

This section discusses our MPTCP implementation. We briefly discuss changes in TCP classes and describe the upstreaming process.

4.1 Why MPTCP into ns-3-dev?

A common problem with all three previous implementations is that they are incompatible with the current ns-3 stack. Because of this limitation, these implementations are unable to use the latest ns-3 models. Our ns-3-dev implementation is current with the ns-3 tree (ns-3-dev) and is compatible with the latest version ns-3.29 as well as potentially compatible with the future ns-3 versions.

4.2 Changes in TCP Stack of ns-3

As described earlier, the ns-3 TCP stack is rapidly evolving. The ns-3 TCP stack was *rewritten* in ns-3.8 and later *refactored* in ns-3.25 resulting in major changes and modifications to several functions and variables in TCP classes. There are many noticeable enhancements to TCP classes. Several functions have been modified and although they have the same name as in previous versions of ns-3 they contain different parameters.The TcpSocketBase class has had major changes over the evolution of ns-3. The class no longer handles congestion control and new congestion control classes that are subclasses of the *TcpCongestionOps* class have been introduced. The TcpSocketState class keeps track of the congestion state of a connection. Congestion control related variables such as m_highTxMark, m_nextTxSequence, etc., have been moved from the TcpSocketBase class into the TcpSocketState class. Furthermore, new variables for congestion control such as m_bytesInFlight, m_pacing, m_rcvTimestampValue, m_lastRtt, etc., have been introduced in the TcpSocketState class. Several variables have been added to the TcpSocketBase class such as m_highTxAck, m_bytesInFlight, m_dataRetrCount, m_dataRetries, m_sndScaleFactor, m_rcvScaleFactor, etc. Furthermore, some functions have been removed from TcpSocketBase such as FirstUnacked-Seq(), GetRttEstimator(), SendEmptyPacket (TcpHeader& header), UpdateTxBuffer(), etc. Whereas, several new functions have been added such as UpdateRttHistory(), UpdateCwndInfl, LimitedTransmit (), FastRetransmit (), etc. Selective Acknowledgments (SACK) and Explicit Congestion Notification (ECN) have also been incorporated into the TCP stack in ns-3. The TcpTxBuffer class has been updated in accordance with RFCs and the Linux operating system to implement mechanisms for SACK and management of bytes in flight.

4.3   Upstreaming Process

We used Coudron et al. implementation [14] as base code and upstreamed it into
the current ns-3-dev by making the necessary modifications to relevant classes.
During the process, we have tried to minimize dependencies of MPTCP classes
i.e., MpTcpSocketBase, MpTcpSubflow, TcpOptionMpTcp, etc. on TCP classes.
We have also made appropriate changes to the TcpL4Protocol, TcpRxBuffer and
TcpSocketBase classes to make existing MPTCP code compatible with ns-3-dev.
While compiling MPTCP code we faced two types of errors: *compiler related
errors* and *incompatibility* related errors with the TCP stack.

4.4   Path Management Implementation

Path management is a critical functionality of MPTCP and was missing in the
previous ns-3.23 implementation. We implement a path management component to initiate subflows for three path managers default, ndiffPorts and
fullmesh as described in Sect. 2.1. New classes MpTcpNdiffPorts and MpTcp-
FullMesh were created for ndiffports and fullmesh path manager. Figure 5
describes the path management component of MPTCP. When a client receives
the first DSS ACK it initiates path managers as described in RFC 6824 [21]. The
user can configure a path manager prior to simulation. By selecting the ndiffports path manager one can control the number of subflows through a socket
API MaxSubflows. Similarly, the fullmesh path manager creates a full-mesh of
subflows amongst all available IP addresses at the sender and receiver.

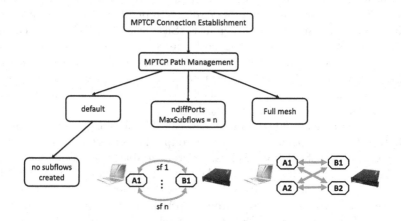

Fig. 5. Path managers functional diagram

5   Simulation

In order to evaluate the efficacy of our implementation, we compare simulation
results with Coudron et al. [14]. We simulate a scenario where a multihomed

client is connected to a Wide Area Network (WAN) with two network interfaces i.e. *Ethernet* and *wi-fi*. The client connects to routers with links having a bandwidth of 2 Mbps each while, all other links have a bandwidth of 2.4 Gbps as shown in Fig. 6. We enable per packet ECMP to route packets through the network.

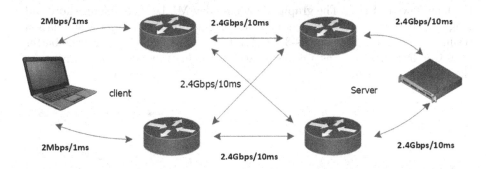

Fig. 6. Simulation topology

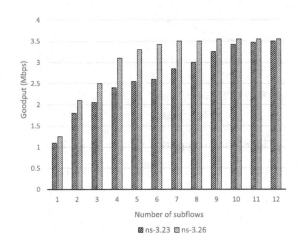

Fig. 7. Goodput for long flows

The client sends a large file to the server and we record the bytes received by the server for different number of MPTCP subflows. We then plot the *goodput* achieved by the MPTCP connection with different number of subflows as shown in Fig. 7. It is evident that with an increase in the number of MPTCP subflows the achieved goodput increases both with ns-3-dev and Coudron et al. [14] implementations. However, the ns-3-dev MPTCP achieves better goodput for less than 8 subflows and this can be attributed to TCP's fast retransmit and fast recovery as well as other enhancements made in ns-3. Notice that, the goodput saturates to approximately to 4 Mbps. Plain TCP performs poorly whereas, increasing the number of subflows allows better utilization of the available capacity. The results are in the same vein as [13]. Furthermore, we perform simulations for *short* and *long flows* for the same topology shown in Fig. 6. We create 4 subflows with ndiffports path manager for each MPTCP connection to complete these flows. We plot the flow completion times (FCTs) for the short flows in Fig. 8 and for the long flows in Fig. 9. The graphs show ns-3-dev MPTCP completes short and long flows earlier than Coudron et al.'s implementation [14]. From the simulation results, it is evident that our ns-3-dev MPTCP implementation performs better than Coudron et al. [14] as a consequence of new features and enhancements in the TCP stack of ns-3.

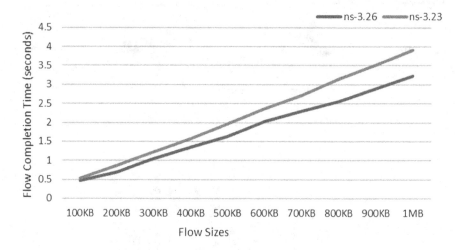

Fig. 8. Flow completion time for short flows

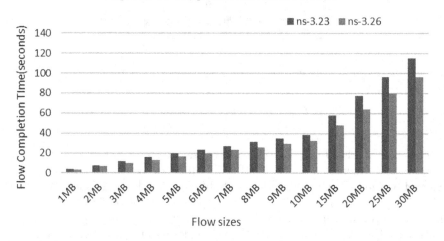

Fig. 9. Flow completion time for long flows

6 Conclusions and Future Work

Simulation tools such as ns-3 are of great help to researchers and network engineers because of the limited access to live network resources and/or testbeds. Presently, ns-3 provides an excellent simulation platform for network engineers and researchers. MPTCP has not been incorporated in the main ns-3 tree by ns-3 community thus far. Efforts have been made in the past to build MPTCP in ns-3 [22,23] and [14]. We build on these implementations, expanding their scope and developing an implementation compatible with the latest ns-3 TCP stack. We implement MPTCP in ns-3-dev and introduce path managers to initiate subflows. We create an MPTCP patch [3] for integration into the main ns-3 tree. Our implementation will help the research community to test new MPTCP proposals and improve upon its functionality and performance. We aim to incorporate our implementation in the ns-3 main tree and additionally develop support for MPTCP congestion control algorithms.

Acknowledgements. We would like to thank Matthieu Coudron for valuable discussions and assistance in navigating through the ns-3 development effort.

References

1. ns-3 website. https://www.nsnam.org/. Accessed 05 Nov 2018
2. ns-3 mptcp patch. https://codereview.appspot.com/369810043/. Accessed 31 Aug 2018
3. ns-3-dev mptcp. https://github.com/Kashif-Nadeem/ns-3-dev-git. Accessed 31 Aug 2018
4. Postel, J.: Transmission Control Protocol. Internet Requests for Comments (1981). https://www.rfc-editor.org/rfc/rfc793.txt
5. Alizadeh, M., et al.: Data center TCP (DCTCP). In: SIGCOMM (2010)
6. Vamanan, B., Hasan, J., Vijaykumar, T.: Deadline aware data center TCP (D2TCP). In: SIGCOMM (2012)
7. Bai, W., et al.: PIAS: practical information-agnostic flow scheduling for datacenter network. In: HotNets (2014)
8. Wischik, D., Handley, M., Bagnulo, M.: The resource pooling principle. ACM SIGCOMM CCR **38**(5), 47–52 (2008)
9. Ong, L., Yoakum, J.: An introduction to the stream control transmission protocol (SCTP). Internet Requests for Comments (2002). https://www.ietf.org/rfc/rfc3286.txt
10. Hasegawa, Y., Yamaguchi, I., Hama, T., Shimonishi, H., Murase., T.: Improved data distribution for multipath TCP communication. In: IEEE Globecom (2005)
11. Zhang, M., Lai, J., Krishnamurthy, A., Peterson, L., Wang, R.: A transport layer approach for improving end-to-end performance and robustness using redundant paths. In: USENIX (ATEC 2004) (2004)
12. Hopps, C.: Analysis of an Equal-Cost Multi-Path Algorithm. Internet Requests for Comments (2000). https://www.rfc-editor.org/rfc/rfc2992.txt
13. Raiciu, C., Barre, S., Pluntke, C., Greenhalgh, A., Wischik, D., Handley, M.: Improving datacenter performance and robustness with multipath TCP. In: ACM SIGCOMM (2011)

14. Coudron, M., Secci, S.: An implementation of multipath TCP in ns3. Comput. Netw. **116**, 1–11 (2017). https://doi.org/10.1016/j.comnet.2017.02.002
15. Ford, A., Raiciu, C., Handley, M., Barre, S., Iyengar, J.: Architectural guidelines for multipath TCP development. Internet Requests for Comments (2011). https://tools.ietf.org/html/rfc6182
16. Carpenter, B., Brim, S.: Middleboxes: taxonomy and issues. Internet Requests for Comments (2001). https://www.rfc-editor.org/rfc/rfc3234.txt
17. Scharf, M., Ford, A.: MPTCP application interface considerations. Internet Requests for Comments (2013). https://tools.ietf.org/html/rfc6897
18. Ford, B., Iyengar, J.: Breaking up the transport logjam. In: ACM HotNets (2008)
19. Multipath Tcp Linux Kernel Implementation. http://multipath-tcp.org/pmwiki.php/Users/ConfigureMPTCP. Accessed 31 Aug 2018
20. Boccassi, L., Fayed, M., Marina, M.: Binder: a system to aggregate multiple internet gateways in community networks. In: LCDNet 2013 (2013)
21. Ford, A., Raiciu, C., Handley, M., Bonaventure, O.: TCP extensions for multipath operation with multiple addresses. Internet Requests for Comments (2013). https://tools.ietf.org/html/rfc6824
22. Chihani, B., Collange, D.: Towards monolingual programming environments. In: WNS3 (2011)
23. Kheirkhah, M., Wakeman, I., Parisis, G.: Multipath-TCP in ns-3 (2015). https://arxiv.org/abs/1510.07721v1
24. Barré, S., Paasch, C., Bonaventure, O.: MultiPath TCP: from theory to practice. In: Domingo-Pascual, J., Manzoni, P., Palazzo, S., Pont, A., Scoglio, C. (eds.) NETWORKING 2011. LNCS, vol. 6640, pp. 444–457. Springer, Heidelberg (2011). https://doi.org/10.1007/978-3-642-20757-0_35
25. Allman, M., Paxson, V., Blanton, E.: TCP congestion control. Internet Requests for Comments (2009). https://www.rfc-editor.org/rfc/rfc5681.txt

A Novel Security Framework for Industrial IoT Based on ISA 100.11a

Hyunjin Kim, Sungjin Kim, Sungmoon Kwon, Wooyeon Jo,
and Taeshik Shon[(⊠)]

Department of Computer Engineering, Ajou University, Suwon, Korea
hyunjin.infosec@gmail.com,
{ksjskyblue,tsshon}@ajou.ac.kr,
calmcombat@gmail.com, dndusdndus12@gmail.com

Abstract. This paper proposes a security assurance technology of IoT devices using their relevant standard, focusing on ISA100.11a, one of the ICS wireless communication protocols. The proposed security assurance technology is divided broadly into communication test and security function assessment. In detail, the communication test is divided into baseline operation test, resource robustness testing, and packet manipulation testing. The security function assessment conducted with the devices that have passed communication testing is proposed differing the required items, divided by the components of ISA100.11a, such as a field device, backbone router, and host so that an assessment appropriate for the hardware specifications and roles of each component is achieved. In addition, the paper seeks to facilitate the implementation and application of the proposed security assurance technology by proposing concrete methods or criteria for communication testing and security function assessment. Finally, this paper attempts to verify the conformance of the proposed security assurance by testing the security assurance technology in a testbed with a network environment where the standard ISA100.11a can work network environment.

Keywords: Industrial Control System (ICS) · Industrial IoT (IIoT) ·
ISA100.11a · Security framework

1 Introduction

The existing Industrial Control System (ICS) attempted to establish the reliability of security based on the isolated network through the separation from the external network, but the malicious codes, which had attacked the targets of ICS-related companies and institutions such as Stuxnet (2010), Duqu (2011), Flame (2012), Gauss (2012), Shamoon (2012), Havex (2014) and Black Energy (2014), was proven that there is a security vulnerability in the isolated network environment. Moreover, the Industrial Internet of Things (IIoT), that apply the Internet of Things (IoT) technology to existing ICS, is introduced and deployed for increasing service economically and efficiently. It means that the existing isolated network gradually gets openness and inherits security vulnerability in the existing Information & Communication Technology (ICT) environment.

© ICST Institute for Computer Sciences, Social Informatics and Telecommunications Engineering 2019
Published by Springer Nature Switzerland AG 2019. All Rights Reserved
T. Q. Duong et al. (Eds.): Qshine 2018, LNICST 272, pp. 61–72, 2019.
https://doi.org/10.1007/978-3-030-14413-5_5

The security vulnerabilities of ICS are increasing and can cause serious damage to economic, social, and human life, security must be considered before deploying and operating the IIoT technology in ICS.

One of the major technologies of the IIoT is the wireless communication technology, it has many advantages such as the convenience of maintenance, cost savings, scalability, interoperability, and mobility, so IIoT has been actively applied to the industrial field. According to the 2016 HMS, wireless network takes 4% in the entire market of the industrial network, but it reports that the Compound Annual Growth Rate (CAGR) amounts to 30%, so it is noted that the application of the wireless network is becoming gradually more active in the industry. For deploying and operating the wireless communication in the real industrial environment, ZigBee, WirelessHART, and ISA100.11a protocols are in the limelight, which are based on IEEE.802.15.4 Standard that has advantages in the node price and the number of nodes supported and supports low power consumption and low processing capability. Of them, ISA100.11a is a protocol with many benefits such as time management, security, interoperability and the use of open standards. But ISA100.11a was enacted lately, there are insufficient related studies and assurance systems. Therefore, this study would promote the improvement of the security of the ICS based on ISA100.11a by proposing a security assurance framework for a system using it.

First, Sect. 2 examines the overall background and present condition, such as the wireless communication network background of the ICS, representative protocols of the relevant wireless communication network and the present condition of the assurance systems of the ICS. Section 3 proposes a security framework for the ICS based on ISA100.11a. Section 4 facilitates the implementation and application of the security framework in the ICS, suggesting methods for measuring individual components of the proposed security framework and criteria for assessment and also execute framework in a testbed network to which ISA100.11a can be applied and carries out the proposed security assurance framework for the target devices. And lastly, Sect. 5 draws conclusions and introduces the follow-up studies.

2 Background

The existing ICS was mostly wire communication-based system using industrial serial communication protocols such as PROFIBUS, Modbus and CAN and industrial ethernet communication protocols such as EtherNet/IP, PROFINET and EtherCAT because of strict conditions of the characteristics of the environment of the industry, such as real-time communication, time-limited processing, high availability, functional safety and security (see Fig. 1). But nowadays wireless communication is deployed in ICS environment for reducing maintenance cost and increasing the usability of systems. To use wireless communication while satisfying requirements of the environment of the industry, ICS groups developed various technology and specification, such as Zigbee, WirelessHART and ISA100.11a [1, 2]. Among these standards, ISA100.11 was developed most recently and use open standard from physical layer to transport layer.

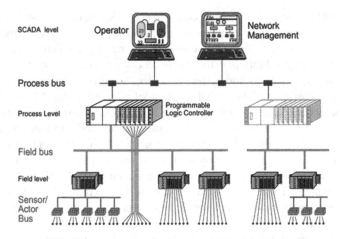

Fig. 1. Location of the field bus in the plant hierarchy [3]

ISA100.11a is a wireless communication network standard issued by the International Society of Automation (ISA), a non-profit organization in the U.S., in order to overcome the disadvantage of the wire communication protocol in the existing ICS. Reliability, security, robustness, quality, interoperability, existing system, compatibility and large network support were considered requirements, and the main characteristics include low power/low speed wireless network use, compatibility with other communication standards, open standard and IPv6 support. In 2009, ISA100.11a was officially announced, and it was approved as IEC 62734 Standard at the International Electrotechnical Commission (IEC) in 2011. The network components of ISA100.11a include adapter, gateway, network manager, security manager, handheld and field device, and it can perform communication by the composition (See Fig. 2).

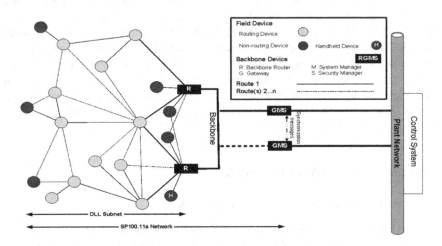

Fig. 2. ISA100.11a network configuration [4]

The ISA100.11a assurance test is conducted by ISA and two tests can be performed. The first test is stack test that ensures ISA100 wireless stack conformance prior to device interoperability testing and second test is device test that ensures ISA100 wireless device interoperability in native mode. In order to conduct two tests of Device Under Test (DUT), a gateway and system manager can be provided, which constitutes an ISA100.11a network, and an official test tool kit certified by ISA100 WCI is used. The entire test process is performed according to the ISA100 wireless communication specification. The products that have conformance and interoperability with ISA100.11a obtain a certificate. Like this, the assurance of ISA100.11a turned out to be the first step which is mostly formed by the tests on protocol stack and interoperability, and tests on stack and interoperability are performed on 6LoWPAN, one of the main referenced standards. Therefore, in order to assure the security of the ICS that uses ISA100.11a, it is necessary to draw additional security requirements and propose a security framework.

Some studies are conducted about the framework and methodology of a security vulnerability study and security test of the main protocols of ISA100.11a, 6LoWPAN and IPv6 and apply the result to the security assurance framework of this study. Studies analyzing the security vulnerability of 6LoWPAN include Le [5] and Redwan [6]. Anhtuan Le analyzed security threats that might occur in a 6LoWPAN network environment and suggested security countermeasures based on Intrusion Detection System (IDS). The study examined investigated OSI 7layers security threat in the existing WSN environment and security vulnerability that might occur in a known 6LoWPAN network and designed an IDS that could respond to a Quality of Service (QoS) attack abusing the vulnerabilities. Hassen Redwan proposed an end-to-end authentication protocol to detect a fragmentation attack method possible on a 6LoW-PAN network and defend the attack. The security vulnerability of 6LoWPAN drawn from the studies is based on the wireless communication vulnerability generally known, which attacks on the integrity, availability and confidentiality of a network from eavesdropping, intercepting and manipulating packets.

The representative standard documents and guideline documents related to the methodology and evaluation of security testing include of NIST SP 800-42, SP 800-53A and SP 800-115, ISECOM OSSTMM, OISSG ISSAF and OWASP Group Testing Guide v5. These documents specify that security should be evaluated through a test, examination and interview and suggest that each method should be performed through the steps such as plan, execution and post-execution. This study proposed penetration testing related to test and examination and a security authentication framework that evaluates security through examination and analysis of the objects or functions of the target.

3   Proposed Security Frameworks

3.1   Overview of Proposed Security Framework

Similar to the Achilles certification and ISASecure certification program, which give certification about embedded device by testing procedure, this section proposed security assurance framework for device of industrial control system based on ISA100.11a protocol. This security assurance framework consists of two main categories. One is communication test; the other is security function test. All devices that use to ISA100.11a protocol must be satisfied with same communication test, but components of ISA100.11a network, such as field device, router and host, are satisfied with different number of security functions. Communication test is specific test about ISA100.11a communication and focus on transport and network protocol layer in ISA100.11a protocol. It is helpful to test other protocols that are based on these layers and to apply common test item to various component devices of ISA100.11a network (See Fig. 3).

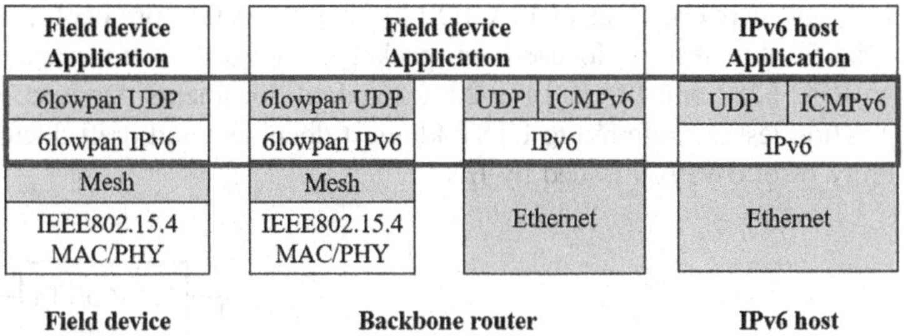

Field device Application
6lowpan UDP
6lowpan IPv6
Mesh
IEEE802.15.4 MAC/PHY
Field device Application
6lowpan UDP
6lowpan IPv6
Mesh
IEEE802.15.4 MAC/PHY
UDP
ICMPv6
IPv6
Ethernet
IPv6 host Application
UDP
ICMPv6
IPv6
Ethernet
Field device
Backbone router
IPv6 host

Fig. 3. Protocol stack of network component ISA

Security function test is assurance procedure to assess whether ISA100.11a devices are satisfied with requirement of security functions. As each component of ISA100.11a network, such as field device, router and host, have different purpose and H/W specification, different number of requirements are suggested for each component. Organization of proposed security assurance framework are shown in the following Fig. 4.

Communication Evaluation
Baseline operation Test
Resource Robustness Test
Malicious Packet Test
Functional Security Assessment
Manager
Router
Field Device

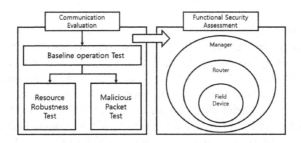

Fig. 4. Proposed security framework

3.2 Communication Test

Communication test is consisting of three test components: baseline operation test, resource robustness test, packet manipulation test. These tests are focused on network layer and transport layer of ISA100.11a communication protocol and are confirmed about maintaining normal operation of target device during test procedure. If the target device occurs following abnormal operation, it means that target device is not satisfied with related test item.

– **Program instability:** program crash or restart when test is being executed.
– **Suspension of network operation:** deny response of application program when test is being executed.
– **Abnormal exhaustion of resource:** occur abnormal CPU availability or memory leak when test is being executed.

Baseline Operation Test

Baseline operation test is basic test of packet creation and processing that is based on transport layer and network layer of ISA100.11a communication protocol. Similar to following Fig. 5, this test is focused on packet compression and fragmentation according to RFC 6282 and ISA100.11a-2010 standard documents. Derived items of baseline operation test are referred to ETST Plugtest document and draft document of Interoperability of 6LoWPAN issued by IETF.

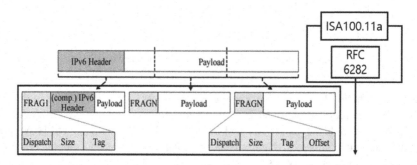

Fig. 5. Scope of baseline operation test and reference standards

Resource Robustness Test

Resource robustness test perform consumption of specific resources in target device and evaluate processing and protection capability of target device under transmission of high speed and flooding packet. This test evaluates that existing communication connection is continuously operating within the time limit. So that it assurance that target device is maintaining essential service under exhaust resource attack. Derived items of resource robustness test are referred to level 1 of Archilles communication certification that is assurance of resource robustness and are transformed into suitable ISA100.11a.

Packet Manipulation Test

Packet manipulation test evaluate processing and protection capability of target device by creating and transmitting of abnormal packet, such as non-conformed standard packet, invalid value packet, invalid sequence packet and known attack packet. This test can use fuzzing methodology (Blackbox test) and manual methodology (Whitebox test) for generating and transferring attack packet. Target device not only does not become program instability, suspension of network operation and abnormal exhaustion of resource but also executes protection action under packet manipulation test. Derived items of packet manipulation test are referred to IEC 62443 and RFC standard.

3.3 Security Function Testing

Security Function testing confirm operating and implementing of security functions for maintaining and operating target device. This test assures target device by checking function name or command of OS. Security Function testing are needed to propose appropriate requirement of security function depending on different hardware specifications and roles of device. This study divides devices of ISA100.11a network into three components: a field device, backbone router, and host. And then, it proposes three different evaluation items of security function for each component. For example, a field device is satisfied with assurance level 1 and backbone router is satisfied with assurance level 1 & 2 and host is satisfied with assurance level 1 & 2 & 3. Derived items of Security Function testing are referred to NIST SP800-53, NIST SP 800-52, NERC CIP, PP (Protection profile) for WLAN access systems/client and PP profile of an industrial wireless base-station. And then, these items are adopted about different level of assurance method in IEC 62443-4-2.

4 Measurement and Evaluation Method

Specifying evaluation method and environment about proposed items is important methodology for proving objectivity, consistency, validity and reliability in assurance program. Furthermore, this can be useful to use self-testing for finding and making up for security vulnerability in target device and to improve overall security level about related device. This section specifies method of evaluation and measurement about proposed security assurance framework and then experiment with testbed that is able to apply ISA100.11a network.

4.1 Measurement Method for Communication and Security Function Test

The ISA100.11a protocol is based on IEEE 802.15.4 standard in the physical layer and data link layer, but it defines IEEE 802.15.4 expanded for channel hopping and mesh network in the data link layer. In the network and transport layers, it uses open standard, 6LoWPAN connected to the UDP and IPv6 standard. Therefore, communication testing method and configuration are based on RFC standard document.

Baseline Operation Test
Baseline operation test is conformed about generating and processing 6LoWPAN complying with RFC standard and about derived items by analyzing response packet for transferred echo packet to target device. Configuration and process of baseline operation test is shown in Fig. 6.

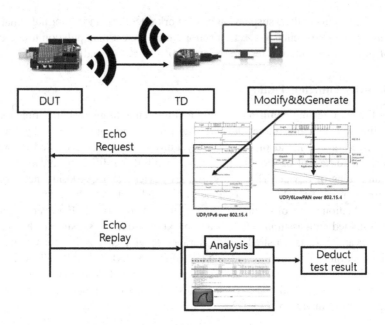

Fig. 6. Configuration and process of baseline operation test

Resource Robustness Test

Resource robustness test is exhausting specific resource of target device by transferring flood packets and evaluating capability of processing and protecting about target device. The test is checked about time interval of existing communication connection and is monitored about operation status of target device under the test. Configuration and process of resource robustness test is shown in Fig. 7.

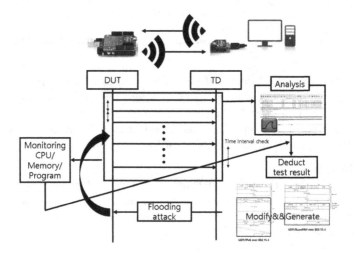

Fig. 7. Configuration and process of resource robustness test

Packet Manipulation Test

The test is checked about capability of processing and protecting in target device by transferring non-conformed standard packet, invalid value packet, invalid sequence packet and known attack packet. Configuration and process of packet manipulation test is shown in blow Fig. 8.

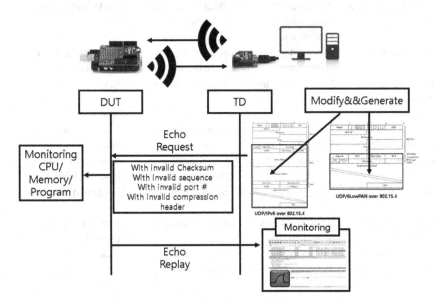

Fig. 8. Configuration and process of packet manipulation test.

Security Function Test

Security Function testing confirm operating and implementing of security functions for maintaining and operating target device. This test assures target device by checking function name or command of OS. However, same security function has various function name or command according to target OS. So, the following Table 1 briefly guides example checklist about security function test.

Table 1. Example checklist of security function test

Checklist of security function			
Role-based access	User password management	System use notification	Audit record generation
Dual authentication access	Monitor unsuccessful login attempt	Local session locking timeout	Audit record time-stamp
Least privilege default access	Record successful login	Remote session termination timeout	Non-repudiation of audit record

(continued)

Table 1. (*continued*)

Checklist of security function			
Administrator role	Previous login notification	Monitoring unauthorized connection	Additional content of audit record
Administrator access support	Password modification notification	Disable wireless networking	Audit fault warning
Write protection	Password strength enforcement	Basic device authentication	Basic protection of audit record
Basic protection of executable code	Unsuccessful login attempt	Session creation	Cryptographic protection of audit record
Crypto protection of executable code	Password minimum strength	Basic protection of session	Audit record of system area
Basic protection of OS	Password encryption	Cryptographic protection of session	maintenance of essential service
Crypto protection of OS	Encrypted password protection	Session termination	Backup system support
Security function verification	Basic confidentiality of system data	Session timeout	Recovery system support
Security function isolation	Cryptographic mechanism	Information flow enforcement	Incident response support

4.2   Experiment Methodology of Proposed Framework

For experiment of proposed security assurance framework, it was configured to a network environment where the standard ISA100.11a can work network environment. By using arduino Arduino XBee Artenna series 1 (802.15.4 1 mW) and arduino uno R3 board (Fig. 9).

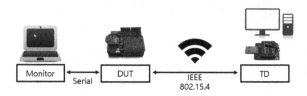

Fig. 9. Experimental configuration of proposed framework

And then for configuration of UDP/6LoWPAN stack on this H/W components, it was used to pIPv6 source code, which implement UDP/6LoWPAN for arduino uno based on Contiki OS [7, 8] (Fig. 10).

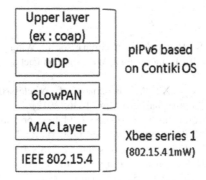

Fig. 10. Experimental configuration of protocol stack

5 Conclusion

ISA100.11a, one of the main wireless communication protocols in industrial environment, has high scalability, reliability, security, robustness, quality, interoperability, compatibility of existing system and large network support. Therefore, it is actively researching and applying in industry site. Due to wireless communication system characteristic using ICT based on main infrastructure, it is necessary to security technology and security assurance program about cybersecurity threats. However, there are lacking studies to test and evaluate the security of the wireless communication technology for ICS. Therefore, this study proposed a security assurance technology of the devices focusing on ISA100.11a, one of the wireless communication protocols for ICS. The proposed security assurance technology was divided broadly into communication testing and security function assessment, and the communication testing was divided into baseline operation testing, resource robustness testing, and packet manipulation testing. A security function assessment conducted with the devices that have passed communication testing was proposed differing the required items, divided by the components of ISA100.11a, such as a field device, backbone router, and host so that an assessment appropriate for the hardware specifications and roles of each component is achieved. In addition, this study seeks to facilitate the implementation and application of the proposed security assurance technology by proposing concrete methods or criteria for communication testing and security function assessment.

Acknowledgments
- This research was supported by the MSIT (Ministry of Science and ICT), Korea, under the ITRC (Information Technology Research Center) support program (IITP-2018-2016-0-00304) supervised by the IITP (Institute for Information & communications Technology Promotion).
- This research was supported by Basic Science Research Program through the National Research Foundation of Korea (NRF) funded by the Ministry of Science, ICT & Future Planning (NRF-2018R1D1A1B07043349).
- This work was supported by an IITP grant funded by the Korean government (MSIT) (No. 2018-0-00336, Advanced Manufacturing Process Anomaly Detection to prevent the Smart Factory Operation Failure by Cyber Attacks).

References

1. Lennvall, T., Svensson, S., Hekland, F.: A comparison of WirelessHART and ZigBee for industrial applications. In: IEEE International Workshop on Factory Communication Systems (WFCS 2008) (2008)
2. Nixon, M., Round Rock, T.X.: A comparison of WirelessHART and ISA100.11a. Whitepaper, Emerson Process Management, pp. 1–36 (2012)
3. Kirrman, H.: Industrial communication systems-field bus: principles. https://web.fe.up.pt/ ~asousa/sind/acetat/AI_EPFL/AI_3xx_Field_bus_OSI_MVB.pdf. Accessed 17 Oct 2018
4. Analysis of wireless industrial automation standards: ISA-100.11a and WirelessHART. https://www.isa.org/standards-publications/isa-publications/intech-magazine/2012/december/ web-exclusive-analysis-wireless-industrial-automation-standards-isa-100-11a-wirelesshart/. Accessed 17 Oct 2018
5. Le, A., Loo, J., Lasebae, A., Aiash, M., Luo, Y.: 6LoWPAN: a study on QoS security threats and countermeasures using intrusion detection system approach. Int. J. Commun. Syst. 25(9), 1189–1212 (2012)
6. Redwan, H., et al.: SAKES: secure authentication and key establishment scheme for M2M communication in the IP-based wireless sensor network (6L0WPAN). In: 2013 Fifth International Conference on Ubiquitous and Future Networks (ICUFN). IEEE (2013)
7. A small CoAP implementation for microcontroller. https://github.com/1248/microcoap. Accessed 17 Oct 2018
8. Arduino pico IPv6 stack. https://github.com/telecombretagne/Arduino-pIPv6Stack. Accessed 17 Oct 2018

Social-Aware Caching and Resource Sharing Optimization for Video Delivering in 5G Networks

Minh-Phung Bui[1,2]([✉]), Nguyen-Son Vo[1], Tien-Thanh Nguyen[3],
Quang-Nhat Tran[1,4], and Anh-Tuan Tran[4]

[1] Duy Tan University, Da Nang, Vietnam
buiminhphung@gmail.com, vonguyenson@dtu.edu.vn,
tranquangnhat1983@gmail.com
[2] Van Lang University, Ho Chi Minh City, Vietnam
[3] Quang Binh Department of Science and Technology, Dong Hoi, Vietnam
tienthanh5@gmail.com
[4] Ho Chi Minh City University of Transport, Ho Chi Minh City, Vietnam
anhtuangtvthcm@gmail.com

Abstract. The proliferation demand of mobile users (MUs) for video contents, which will occupy up to 78% of data traffic by 2021, poses a serious challenge of system delivery capacity to the macro base stations (MBSs) and the small cell base stations, e.g., femtocell base stations (FBSs), in 5G networks. In this paper, we propose a social-aware caching and resource sharing (SCS) scheme that can help the MBSs and the FBSs relax the backhaul links and provide the MUs with high system delivery capacity. Particularly, we formulate an SCS optimization problem under the constraints on the number of replicas of each video cached in the FBSs and the target signal to interference plus noise ratio (SINR) of the cellular users (CUs) that share the downlink resources. This problem is then solved for maximum system delivery capacity by finding the best placements to cache the videos in the FBSs and the best device-to-device (D2D) pairs shared the same downlink resources with the CUs to offload the videos over D2D communications. Importantly, the behavior of MUs to access the videos and the social relationship of each D2D pair are considered in the SCS optimization problem to efficiently improve the system performance. Simulation results are shown to demonstrate the benefits of the proposed SCS scheme compared to other conventional schemes.

Keywords: 5G caching · D2D communications ·
Downlink resource sharing · Social-aware networks · Video delivering

1 Introduction

By 2021, there will be 11.6 billion mobile devices connected to wireless networks, generating a huge amount of data traffic, i.e., reaching 49 exabytes per month [1].

© ICST Institute for Computer Sciences, Social Informatics and Telecommunications Engineering 2019
Published by Springer Nature Switzerland AG 2019. All Rights Reserved
T. Q. Duong et al. (Eds.): Qshine 2018, LNICST 272, pp. 73–86, 2019.
https://doi.org/10.1007/978-3-030-14413-5_6

In this scenario, video traffic which yields 78% of data traffic will be challenging for 5G networks to serve mobile users (MUs) high quality of service (QoS). One of the main challenges is the extremely congestion at the backhaul links of the macro base stations (MBSs) and the small cell base stations (SCBs). This in turn degrades the system delivery capacity of 5G networks. To address this challenge, caching techniques, e.g, caching at MBSs, SCBs, and/or MUs, have been feasibly proposed for 5G network, without changing network infrastructure [2].

Caching at the MBSs (MBS caching) is a simple method that can reduce the backhaul traffic, while providing the MUs with high QoS [3,4]. However, it is not high efficient enough for relaxing the backhaul links under a massive number of MUs. To further assist MBS caching, caching at SCBs, e.g., femtocaching, has been studied to gain high system capacity and low latency [5–8]. The congestions at the backhaul links of the MBSs and SCBs are also reduced by edge caching technique, namely device-to-device (D2D) caching [9–11]. In addition, the most efficient caching technique that has been carefully studied is multi-tier caching. Multi-tier caching enables to cache at all MBSs, SCBs, and MUs simultaneously to reduce the traffic and the energy consumption at the MBSs [12], increase the system capacity [13], and deliver the videos to the MUs efficiently [14,15].

Importantly, it is certain that if the impact of social relationship between the MUs, following the Indian Buffet Model [16,17], is taken in to account, the performance of caching techniques is significantly improved. In particular, based on the social-tier factor, the MUs who are of similar interests, enough encounter duration, and adjacent to each other, will communicate with each other via D2D communications. The social-tier factor together with video request probability and distance of D2D pairs obtained from practical cellular networks can be also exploited to increase the system throughput by caching in D2D networks [18–20].

Motivated by the aforementioned analysis, in this paper, we propose an optimal social-aware caching and resource sharing (SCS) solution that can help the MBSs and the femtocell base stations (FBSs) relax the backhaul links and provide the MUs with maximum system delivery capacity. To do so, we take the advantages of social relationship of D2D pairs and the video popularity to find both the optimal caching placements at the FBSs and the optimal selections of each MU (namely cellular user (CU) that share its downlink resource) and the D2D pairs (that benefit from the downlink resource shared by the CU). We further consider the target peak signal-to-noise ratio (PSNR) of the CUs to limit the effect of the interference generated by D2D communications on the CUs, and thus guaranteeing a high QoS for the CUs.

The rest of this paper is organized as follows. In Sect. 2, we introduce the system models consisting of 5G SCS, channel, social, and system delivery capacity models. Based on the system models, the SCS problem is formulated and solved in Sect. 3. We present the performance evaluation in Sect. 4. Finally, Sect. 5 concludes the paper.

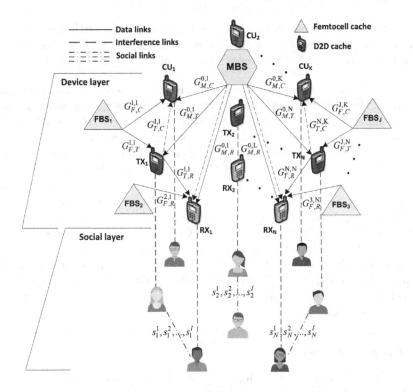

Fig. 1. 5G social-aware caching and downlink resource sharing model.

2   System Models

In this section, we first propose the 5G SCS model and describe how it works. Then, we present the channel models between the MBS and the MUs, the FBSs and the MUs, and the TXs and the RXs. Afterward, the social model of each D2D pair is introduced. Finally, the system delivery capacity is derived as the objective function of the SCS optimization problem.

2.1   5G SCS Model

In this paper, we consider a 5G network of two layers, i.e., device layer and social layer, as shown in Fig. 1. The device layer consists of one MBS, J FBSs, $(K + 2N)$ MUs, and I videos. There are three types of MUs including K cellular users (CUs) and N D2D pairs. Each D2D pair has a D2D transmitter (TX) and a D2D receiver (RX). The CUs share their downlink resources with the D2D pairs for D2D communications. In this network, an MU can request the videos regularly from the MBS, the FBSs, or the TXs with videos cached in advance. It means that besides being served by the MBS, the joint femtocaching and D2D caching scheme is deployed to further serve the MUs higher system delivery capacity. Assuming that we consider a scenario where the network parameters remain at least in a maximum video streaming session in a particular area, e.g., stadiums, concert or meeting halls, campuses, and office buildings, the detailed SCS is deployed at the MBS in three steps as follows:

- Step 1 - Updating network parameters: If the network has any significant changes, the MBS updates the new system parameters, e.g., number of videos (I), number of CUs (K), number of D2D pairs (N), social relationship of each D2D pair, system bandwidth, and information of channels, etc.
- Step 2 - Maximizing system delivering capacity: Based on the parameters in step 1, the MBS formulates the SCS optimization problem and solves it for optimal caching index $u_{j,i}$, $j = 1, 2, \ldots, J$, $i = 1, 2, \ldots, I$ and optimal sharing index $v_{k,n}$, $k = 1, 2, \ldots, K$, and $n = 1, 2, \ldots, N$, for maximizing the system delivering capacity. Here, $u_{j,i} = 1$ if the FBS j decides to cache the video i, otherwise $u_{j,i} = 0$ and $v_{k,n} = 1$ if the CU k decides to share its downlink resource with the D2D pair n, otherwise $v_{k,n} = 0$.
- Step 3 - Caching videos and sharing downlink resource: After solving the SCS optimization problem, the MBS assigns which FBS to cache which video and which CU to share its downlink resource to which D2D pairs, for delivering the videos to the MUs.

2.2   Channel Model

For the ease of modelling the channels, the channel splitting and F-ALOHA [21,22] are used to control the cross-tier and co-tier interference due to the overlaid problem of the MBS and the FBSs. A CU can share its downlink resource with any D2D pair of TX and RX. During the resource sharing, the transmissions of the MBS and the TXs have interference effects on the RXs and the CUs, respectively. We denote $G_{S,D}^{s,d}$ as the channel gains between S and D; here S $\in$ {M, F, T} standing for {MBS, FBS, TX} and D $\in$ {C, T, R} standing for {CU, TX, RX}; s $\in$ {j, n}, $j = 1, 2, \ldots, J$ except that $j = 0$ indicates the MBS, $n = 1, 2, \ldots, N$ and d $\in$ {k, n}, $k = 1, 2, \ldots, K$. The $G_{S,D}^{s,d}$ is modeled as [22]

$$G_{S,D}^{s,d} = h_{S,D}^{s,d} g_{S,D}^{s,d}, \tag{1}$$

where $h_{S,D}^{s,d}$ is the exponential power fading coefficient and $g_{S,D}^{s,d} = ||h||^{-\xi}$ is the standard power law path loss function in which ξ is the path loss exponent, h is the distance between S and D, and $||.||$ is the Euclidean norm.

2.3   Social Model

We take the social model, i.e., social relationship between the TX and the RX of the D2D pair n, into account to compute the probability that if this pair has a relationship close enough or not, for offloading the video i of duration T_{min}^i. To do so, let X_m be the contact duration of the D2D pair n and X_n be the number of encounters, $m = 1, 2, \ldots, X_n$, the expected contact duration M_n and the variance V_n are sequentially given by [16,17]

$$M_n = \frac{\sum_{m=1}^{X_n} X_m}{X_n} \tag{2}$$

and

$$V_n = \frac{\sum_{m=1}^{X_n} (X_m - M_n)^2}{X_n}. \tag{3}$$

By following [18,23–25], we have the encounter duration distribution modelled as gamma distribution expressed as

$$X \sim \Gamma(\kappa_n, \theta_n) = \Gamma(M_n^2/V_n, V_n/M_n) \tag{4}$$

and the probability density function (PDF) is defined as

$$f(x; \kappa_n, \theta_n) = \frac{1}{\theta_n^{\kappa_n}} \frac{1}{\Gamma(\kappa_n)} x^{\kappa_n - 1} e^{-\frac{x}{\theta_n}}, \tag{5}$$

where $\Gamma(\kappa_n) = \int_0^\infty t^{\kappa_n - 1} e^{-t} dt$.

Thus, the probability that the D2D pair n is qualified to offload the video i of duration T_{min}^i, is given by

$$s_n^i = 1 - \int_0^{\delta T_{min}^i} f(u; \kappa_n, \theta_n) du = 1 - \frac{\gamma(\kappa_n, \frac{\delta T_{min}^i}{\theta_n})}{\Gamma(\kappa_n)}, \tag{6}$$

where $\delta \geq 1$ is added to flexibly adjust the duration of all videos and $\gamma(\kappa_n, \frac{\delta T_{min}^i}{\theta_n}) = \int_0^{\frac{\delta T_{min}^i}{\theta_n}} t^{\kappa_n - 1} e^{-t} dt$.

2.4   System Delivery Capacity

The system delivery capacity is defined as the total throughput delivered from the MBS, FBSs, and TXs to the MUs. The system delivery capacity is computed by analyzing the signal to interference plus noise ratio (SINR) of the channels from the MBS, FBSs, and TXs to the MUs, presented in the sequel.

Capacity Delivered to the CUs: The CU k can share its downlink resource with the D2D pair n and receive the video from the MBS or the FBSs. The SINRs of the channels from the FBS j and the MBS to the k-th CU are respectively given by

$$\gamma_{F,C}^{j,k,i} = \frac{u_{j,i} P_F^j G_{F,C}^{j,k}}{N_0} \tag{7}$$

and

$$\gamma_{M,C}^{0,j,k,i} = \frac{(1 - u_{j,i}) P_M^0 G_{M,C}^{0,k}}{N_0 + \sum_{n=1}^N s_n^i v_{k,n} p_{n,i} P_T^n G_{T,C}^{n,k}}. \tag{8}$$

In (7), if the FBS j decides to cache the video i ($u_{j,i} = 1$), the CU k is served by the FBS j over the channel capacity characterized by the transmission power of the FBS j (P_F^j), the channel gain between the FBS j and the CU k ($G_{\mathsf{F,C}}^{j,k}$), and the power of additive white Gaussian noise (AWGN)(N_0). In (8), otherwise ($u_{j,i} = 0$), the CU k is served by the MBS over the channel capacity characterized by the transmission power of the MBS (P_M^0), the channel gain between the MBS and the CU k ($G_{\mathsf{M,C}}^{0,k}$), the interference affected by the transmission power of the TX n (P_T^n) over the channel gain between it and the CU k ($G_{\mathsf{T,C}}^{n,k}$) if the CU k agrees to share the downlink resource with the D2D pair n ($v_{k,n} = 1$), and the AWGN (N_0). In addition, $p_{n,i}$ is the probability of the TX n to cache the video i, which depends on the access rate (i.e., the popularity) of the video i (r_i) and the percentage of available storage of the TX n (β_n), defined as

$$p_{n,i} = ar_i + b\beta_n, \tag{9}$$

where $a, b \in [0,1]$, $a + b = 1$, and by following Zipf-like distribution [26], the access rate of the video i, which represents the behavior of the MUs toward the video i, is defined as

$$r_i = \frac{i^{-\alpha}}{\sum_{i=1}^{I} i^{-\alpha}}, \tag{10}$$

here $\alpha \geq 0$ represents the skewed access rate among different videos.

By using Shannon-like capacity, given the system bandwidth W, the capacity delivered to the CUs is expressed as

$$R_\mathsf{C} = W \sum_{j=1}^{J} \sum_{k=1}^{K} \sum_{i=1}^{I} r_i \left[\log_2 \left(1 + \gamma_{\mathsf{M,C}}^{0,j,k,i} \right) + \log_2 (1 + \gamma_{\mathsf{F,C}}^{j,k,i}) \right]. \tag{11}$$

Capacity Delivered to the TXs: Because the TXs are not affected by the interference from others, the SINRs of the channels from the FBS j and the MBS to the TX n are simply given by

$$\gamma_{\mathsf{F,T}}^{j,n,i} = \frac{u_{j,i} P_\mathsf{F}^j G_{\mathsf{F,T}}^{j,n}}{N_0} \tag{12}$$

and

$$\gamma_{\mathsf{M,T}}^{0,j,n,i} = \frac{(1 - u_{j,i}) P_\mathsf{M}^0 G_{\mathsf{M,T}}^{0,n}}{N_0}, \tag{13}$$

where $G_{\mathsf{F,T}}^{j,n}$ and $G_{\mathsf{M,T}}^{0,n}$ are the channel gains from the FBS j and the MBS to the TX n.

Similarly, the capacity delivered from the FBS j and the MBS to the TX n is expressed as

$$R_\mathsf{T} = W \sum_{j=1}^{J}\sum_{n=1}^{N}\sum_{i=1}^{I} r_i \left[\log_2\left(1 + \gamma_{\mathsf{M,T}}^{0,j,n,i}\right) + \log_2(1 + \gamma_{\mathsf{F,T}}^{j,n,i}) \right]. \qquad (14)$$

Capacity Delivered to the RXs: The capacity delivered to the RX n come from not only MBS and the FBS j but also the TX n. The SINRs of the channels from the TX n, the FBS j, and the MBS to the RX n are given in sequence as follows:

$$\gamma_{\mathsf{T,R}}^{n,k,i} = \frac{s_n^i v_{k,n} p_{n,i} P_\mathsf{T}^n G_{\mathsf{T,R}}^{n,n}}{N_0 + P_\mathsf{M}^0 G_{\mathsf{M,R}}^{0,n} + \sum_{l=1,l\neq n}^{N} s_l^i v_{k,l} p_{l,i} P_\mathsf{T}^l G_{\mathsf{T,R}}^{l,l}}, \qquad (15)$$

$$\gamma_{\mathsf{F,R}}^{j,n,k,i} = \frac{u_{j,i}(1 - s_n^i v_{k,n} p_{n,i}) P_\mathsf{F}^j G_{\mathsf{F,R}}^{j,n}}{N_0}, \qquad (16)$$

and

$$\gamma_{\mathsf{M,R}}^{0,j,n,k,i} = \frac{(1 - u_{j,i})(1 - s_n^i v_{k,n} p_{n,i}) P_\mathsf{M}^0 G_{\mathsf{M,R}}^{0,n}}{N_0}, \qquad (17)$$

where $G_{\mathsf{T,R}}^{n,n}$, $G_{\mathsf{M,R}}^{0,n}$, and $G_{\mathsf{F,R}}^{j,n}$ are the channel gains from the TX n, the MBS, and the FBS j to the RX n, respectively. In (15), the RX n is affected by the interference from not only the MBS but also the others TX $l \neq n, l = 1, 2, \ldots, N$.

So far, the capacity delivered from the MBS, the FBS j, and the TX n to the RX n is respectively expressed as

$$R_\mathsf{R} = W \sum_{n=1}^{N}\sum_{k=1}^{K}\sum_{i=1}^{I} r_i \left[\sum_{j=1}^{J} \left(\log_2(1 + \gamma_{\mathsf{M,R}}^{0,j,n,k,i}) \right. \right.$$
$$\left. \left. + \log_2(1 + \gamma_{\mathsf{F,R}}^{j,n,k,i}) \right) + \log_2(1 + \gamma_{\mathsf{T,R}}^{n,k,i}) \right]. \qquad (18)$$

Finally, from (11), (14), and (18), the overall average system delivery capacity per each MU is given by

$$R = \frac{R_\mathsf{C} + R_\mathsf{T} + R_\mathsf{R}}{K + 2N}. \qquad (19)$$

Solving the SCS optimization problem for maximum R in (19) by finding the optimal caching index $u_{j,i}$ and optimal sharing index $v_{k,n}$ is presented in the following section.

3   SCS Optimization Problem and Solution

To formulate the SCS optimization problem and solve it for maximizing the system delivery capacity R (19) by finding $u_{j,i}$ and $v_{k,n}$, we further consider

Algorithm 1. Exhaustive matrix search

Input: Initial parameters given in Table 1
Output: R^*, $\mathbf{u}_{J \times I}^*$, $\mathbf{v}_{K \times N}^*$
1: Generating two feasible matrix search spaces $\mathcal{U}' \in \mathcal{U}$ and $\mathcal{V}' \in \mathcal{V}$ that satisfy (21)
2: $\mathcal{R} \leftarrow \varnothing$
3: **for** each matrix $u_{J \times I}$ in $\mathcal{U}'$ **do**
4:   **for** each matrix $v_{K \times N}$ in $\mathcal{V}'$ **do**
5:     $R(u_{J \times I}, v_{K \times N}) = R$, computing (19)
6:     $\mathcal{R} \leftarrow \mathcal{R} \cup R(u_{J \times I}, v_{K \times N})$
7:   **end for**
8: **end for**
9: $R^* = \max \mathcal{R}$
10: $\{\mathbf{u}_{J \times I}^*, \mathbf{v}_{K \times N}^*\} = \text{argmax } \mathcal{R}$

---

the constraints on the number of replicas of each video (c_i^*) due to the limited storage capacity of the FBSs and the target SINR of the CUs (γ_0). The SCS optimization problem is expressed as follows:

$$\max_{u_{j,i}, v_{k,n}} R \qquad (20)$$

$$s.t. \begin{cases} \sum_{j=1}^{J} u_{j,i} \leq c_i^*, i = 1, 2, \dots, I \\ \sum_{n=1}^{N} s_n^i v_{k,n} p_{n,i} P_T^n G_{T,C}^{n,k} \leq \dfrac{P_M^0 G_{M,C}^{0,k}}{\gamma_0} - N_0, \\ k = 1, 2, \dots, K, i = 1, 2, \dots, I, \end{cases} \qquad (21)$$

where c_i^* in the first constraint, which is found such that the average number of replicas in the FBSs is maximized for high video hit rate, is given by

$$c_i^* = \underset{\substack{1 \leq c_i \leq J, i = 1, 2, \dots, I \\ \sum_{i=1}^{I} c_i \leq C^*, 1 \leq C^* \leq IJ}}{\arg\max} \sum_{i=1}^{I} r_i c_i, \qquad (22)$$

here C^* is used to limit the number of replicas cached in the FBSs. The linear programming optimization problem (22) can be solved by using primal-dual interior point method [27,28]. In addition, the second constraint of (21) comes from (8) by letting $\gamma_{M,C}^{0,j,k,i} \geq \gamma_0$ and ignoring the term $(1 - u_{j,i})$. It means that the higher value the γ_0 increases, the higher SINR the CUs gain.

It can be observed that finding the optimal caching and sharing indexes, i.e., $u_{j,i}$ and $v_{k,n}$, is equivalent to finding the two optimal matrices $\mathbf{u}_{J \times I}^*$ and $\mathbf{v}_{K \times N}^*$ in the two matrix search spaces: $\mathcal{U} = \{u_{J \times I}^1, u_{J \times I}^2, \dots, u_{J \times I}^{2^{J \times I}}\}$ and $\mathcal{V} = \{v_{K \times N}^1, v_{K \times N}^2, \dots, v_{K \times N}^{2^{K \times N}}\}$, respectively. In this paper, exhaustive matrix search, which is described in Algorithm 1, is used to solve (20) and (21) for $\mathbf{u}_{J \times I}^*$ and $\mathbf{v}_{K \times N}^*$ [15]. The memory and time complexities of the Algorithm 1 are $\mathcal{O}(2^{J \times I + K \times N})$. In case of large scale of 5G networks, it is impractical to search the total space of $2^{J \times I + K \times N}$ matrices done at the MBS, the search space is

Table 1. Parameters setting

Symbols	Specifications
I	4 videos
J	3 FBSs
K	3 CUs
N	5 D2D pairs
$\{\theta_n\}$	$\{20, 20, 20, 20, 20\}$ [16]
$\{\kappa_n\}$	$\{1, 1, 1, 1, 1\}$ [16]
$\{T_{min}^i\}$	$\{15, 20, 10, 5\}$ s
δ	10
α	1
W	5 MHz
P_M^0	5 W
P_F^j	Fixed to 1 W
P_T^n	Fixed to 0.1 W
γ_0	10 dB
N_0	10^{-13} W
ξ	4 (path loss exponent)
$\{\beta_n\}$	$\{0.1, 0.3, 0.5, 0.7, 0.9\}$
a, b	0.5, 0.5
C^*	8

divided into multiple sub-search spaces. An FBS is then assigned a sub-search space for deploying exhaustive matrix search separately to obtain a sub-optimal solution. Finally, all the FBSs send the sub-optimal solutions to the MBS for finding the global optimal solution.

4   Performance Evaluation

In this paper, we simulate the system by deploying the parameters as shown in Table 1. In addition, the distances from the MBS to the MUs, the FBSs to the MUs, the CUs to the TXs, and the TXs to the RXs, are randomly distributed from 100 m to 500 m, 50 m to 250 m, 50 m to 100 m, and 1 m to 50 m, respectively. To evaluate the system performance of our proposed optimization solution (**SCS**), we compare **SCS** to the other three schemes, i.e., none downlink resource sharing (**None-DRS**), average system delivery capacity (**AVE**) and minimum system delivery capacity (**MIN**). In **None-DRS**, there is no downlink resource shared by the CUs; in **AVE**, the system delivery capacity is averaged over the total number of the two feasible matrices generated in the step 1 in the Algorithm 1; and in **MIN**, the minimum system delivery capacity is min $\mathcal{R}$ instead of max $\mathcal{R}$ in the step 9 of the Algorithm 1.

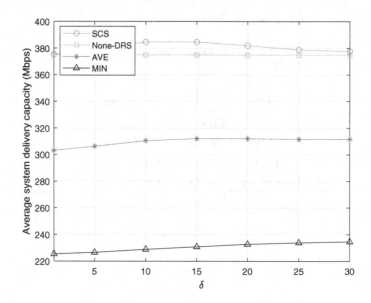

Fig. 2. Capacity performance versus δ.

We first evaluate the performance of **SCS**, **None-DRS**, **AVE**, and **MIN** versus different duration set of all the considered videos by changing δ from 1 to 30. The results in Fig. 2 show that the system delivery capacity increases if we increase δ from 1 to 10, but it decreases if we continue to increase δ. The reason is that if δ is too low (or high), the probability of D2D communications is too high (or low). Both high probability of D2D communications (high interference impact on the CUs) and low probability of D2D communications (not exploiting D2D communications for offloading the videos) result in low system delivery capacity. So, the system delivery capacity gains the highest value at a certain value of δ, i.e., $\delta = 10$. It interestingly means that the duration of videos can be adjusted to meet the social relationship of the D2D pairs, and thus obtaining the highest system delivery capacity. Obviously the performance of **None-DRS** is not affected by δ. In comparison, the proposed **SCS** is better than the others and reduced to the performance of **None-DRS** if δ is too low (or high); and **None-DRS** outperforms **AVE** and **MIN** schemes.

Figure 3 shows the performance of system delivery capacity versus the skewed access rate among different videos by changing α from 0 to 2. We can see that exploiting the skewed access rate improves the system performance. And thus, while **SCS** increases the system delivery capacity with respect to the increase of α, **AVE** and **MIN** decrease it. The results obviously imply that serving less number of high popular videos, i.e., high access rate, yields high system performance. The performance of **None-DRS** is higher than **AVE** and **MIN** and mostly not affected by α due to too less number of videos deployed. And, our

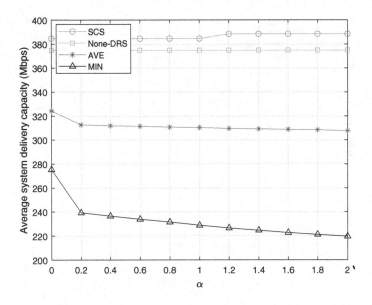

Fig. 3. Capacity performance versus α.

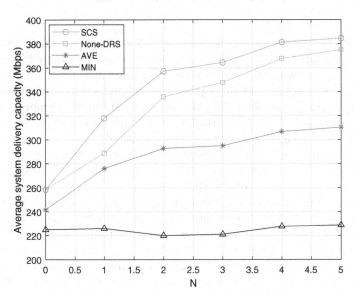

Fig. 4. Capacity performance versus number of D2D pair.

SCS provide higher system delivery capacity compared to **None-DRS**, **AVE**, and **MIN** schemes.

In Fig. 4, the system performance is investigated by changing the number of D2D pairs N from 0 to 5. If $N = 0$, the **SCS** and **None-DRS** have the

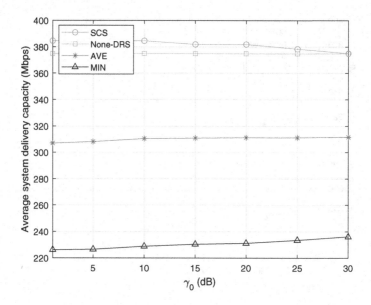

Fig. 5. Capacity performance versus γ_0.

same result because the system delivery capacity comes from the MBS and the FBSs to serve only the CUs. The system delivery capacity clearly increases and becomes saturated with the help of D2D communications when we increase N. The results demonstrate that **SCS** gains the best performance and **None-DRS** is better than **AVE** and **MIN**. In addition, we should select a proper number of D2D pairs such that the PSNR of the CUs is guaranteed and the system delivery capacity is high enough, i.e., before getting saturated.

Finally, we investigate the system performance under the impact of the target SINR of the CUs γ_0. As shown in Fig. 5, the system delivery capacity of **SCS** decreases and approaches **None-DRS** when γ_0 increases. The reason is that if γ_0 is low, more D2D pairs are shared the downlink resource from the CUs to offload the videos for higher system delivery capacity, otherwise, less D2D pairs are for offloading with lower system delivery capacity. However, the system delivery capacity of **MIN** increases because when γ_0 increases, more candidate matrices that cause higher interference impact on the CUs are removed from the search space $\mathcal{V}$. Under the parameters set in Table 1, the decrease of **SCS** and the increase of **MIN** make **AVE** slightly increases. It can be seen obviously that the results of **None-DRS** are not affected by γ_0 because there is no interference impact from D2D communications on the CUs. In this scenario, the proposed **SCS** also surpasses the other three schemes and **None-DRS** outperforms **AVE** and **MIN**.

5 Conclusion

In this paper, we have proposed a social-aware caching and resource sharing optimization solution for video delivering at high capacity in 5G networks. In particular, the social relationship of D2D pairs is exploited to optimally cache the videos in the FBSs and to share the downlink resource of the CUs with the D2D pairs for maximizing the system delivery capacity. The optimization solution is carefully analyzed by taking into account the access rate (i.e., the popularity) of the videos and the target PSNR of the CUs for higher system performance. The interesting result obtained is that a proper duration set of videos selected in accordance with a given set of social relationship of D2D pairs can provide the highest system delivery capacity.

Acknowledgement. This work is supported in part by Newton Prize 2017 and by a Research Environment Links grant, ID 339568416, under the Newton Programme Vietnam partnership. The grant is funded by the UK Department of Business, Energy and Industrial Strategy (BEIS) and delivered by the British Council. For further information, please visit www.newtonfund.ac.uk.

References

1. Cisco Visual Networking Index: Global Mobile Data Traffic Forecast Update, 2016–2021 White Paper - Cisco. https://www.cisco.com/c/en/us/solutions/collateral/service-provider/visual-networking-index-vni/mobile-white-paper-c11-520862.html
2. Li, L., Zhao, G., Blum, R.S.: A survey of caching techniques in cellular networks: research issues and challenges in content placement and delivery strategies. IEEE Commun. Surv. Tutor. **20**, 1710–1732 (2018)
3. Han, W., Liu, A., Lau, V.K.N.: PHY-caching in 5G wireless networks: design and analysis. IEEE Commun. Mag. **54**, 30–36 (2016)
4. Qiao, J., He, Y., Shen, X.: Proactive caching for mobile video streaming in millimeter wave 5G networks. IEEE Trans. Wirel. Commun. **15**, 7187–7198 (2016)
5. Baştuğ, E., Bennis, M., Debbah, M.: Proactive caching in 5G small cell networks. In: Towards 5G: Applications, Requirements and Candidate Technologies. Wiley, Hoboken (2016)
6. Tan, Y., Yuan, Y., Yang, T., Xu, Y., Hu, B.: Femtocaching in wireless video networks: distributed framework based on exact potential game. In: IEEE/CIC International Conference on Communications in China, ICCC 2016 (2016)
7. Kiskani, M.K., Sadjadpour, H.R.: Throughput analysis of decentralized coded content caching in cellular networks. IEEE Trans. Wirel. Commun. **16**, 663–672 (2017)
8. Chen, Y., Ding, M., Li, J., Lin, Z., Mao, G., Hanzo, L.: Probabilistic small-cell caching: performance analysis and optimization. IEEE Trans. Veh. Technol. **66**, 4341–4354 (2016)
9. Chandrasekaran, G., Wang, N., Tafazolli, R.: Caching on the move: towards D2D-based information centric networking for mobile content distribution. In: Proceedings of the Conference on Local Computer Networks, LCN, 26–29 October, pp. 312–320 (2015)

10. Shiroma, T., Nakajima, T., Wu, C., Yoshinaga, T.: A light-weight cooperative caching strategy by D2D content sharing. In: Proceedings of the 5th International Symposium on Computing and Networking, CANDAR, pp. 159–165 (2018)
11. Vo, N.S., Duong, T.Q., Tuan, H.D., Kortun, A.: Optimal video streaming in dense 5G networks with D2D communications. IEEE Access 6, 209–223 (2018)
12. Gregori, M., Gomez-Vilardebo, J., Matamoros, J., Gunduz, D.: Wireless content caching for small cell and D2D networks. IEEE J. Sel. Areas Commun. 34, 1222–1234 (2016)
13. Li, X., Wang, X., Li, K., Han, Z., Leung, V.C.M.: Collaborative multi-tier caching in heterogeneous networks: modeling, analysis, and design. IEEE Trans. Wirel. Commun. 16, 6926–6939 (2017)
14. Lin, P., Song, Q., Yu, Y., Jamalipour, A.: Extensive cooperative caching in D2D integrated cellular networks. IEEE Commun. Lett. 21, 2101–2104 (2017)
15. Vo, N.-S., Duong, T.Q., Guizani, M., Kortun, A.: 5G optimized caching and downlink resource sharing for smart cities. IEEE Access 6, 31457–31468 (2018)
16. Zhang, X., et al.: Information caching strategy for cyber social computing based wireless networks. IEEE Trans. Emerg. Top. Comput. 5, 391–402 (2017)
17. Zhang, Y., Pan, E., Song, L., Saad, W., Dawy, Z., Han, Z.: Social network aware device-to-device communication in wireless networks. IEEE Trans. Wirel. Commun. 14, 177–190 (2015)
18. Zhang, J., Zhang, X., Yan, Z., Li, Y., Wang, W., Zhang, Y.: Social-aware cache information processing for 5G ultra-dense networks. In: 8th IEEE International Conference on Wireless Communications and Signal Processing (WCSP), pp. 1–5 (2016)
19. Ma, C., Ding, M., Chen, H., Lin, Z., Mao, G., Li, X.: Socially aware distributed caching in device-to-device communication networks. In: IEEE Globecom Workshops (GC Wkshps), pp. 1–6 (2016)
20. Ma, C., et al.: Socially aware caching strategy in device-to-device communication networks. IEEE Trans. Veh. Technol. 67, 4615–4629 (2018)
21. Chandrasekhar, V., Andrews, J.G.: Spectrum allocation in tiered cellular networks. IEEE Trans. Commun. 57, 3059–3068 (2009)
22. Cheung, W.C., Quek, T.Q.S., Kountouris, M.: Throughput optimization, spectrum allocation, and access control in two-tier femtocell networks. IEEE J. Sel. Areas Commun. 30, 561–574 (2012)
23. Cha, M., Kwak, H., Rodriguez, P., Ahn, Y.-Y., Moon, S.: I tube, you tube, everybody tubes: analyzing the world's largest user generated content video system. In: Proceedings of the 7th ACM SIGCOMM Conference on Internet Measurement - IMC 2007, pp. 1–14. ACM Press, New York (2007)
24. Benevenuto, F., Rodrigues, T., Cha, M., Almeida, V.: Characterizing user behavior in online social networks. In: Proceedings of the 9th ACM SIGCOMM Conference on Internet Measurement Conference - IMC 2009, New York, USA, pp. 49–62 (2009)
25. Bai, B., Wang, L., Han, Z., Chen, W., Svensson, T.: Caching based socially-aware D2D communications in wireless content delivery networks: a hypergraph framework. IEEE Wirel. Commun. 23, 74–81 (2016)
26. Breslau, L., Cao, P., Fan, L., Phillips, G., Shenker, S.: Web caching and Zipf-like distributions: evidence and implications. In: IEEE INFOCOM 1999, vol. 1, pp. 126–134 (1999)
27. Mehrotra, S.: On the implementation of a primal-dual interior point method. SIAM J. Optim. 2, 575–601 (1992)
28. Zhang, Y.: Solving large-scale linear programs by interior-point methods under the MATLAB environment †. Optim. Methods Softw. 10, 1–31 (1998)

Energy Efficiency in QoS Constrained 60 GHz Millimeter-Wave Ultra-Dense Networks

Huy Thanh Nguyen[1]($\boxtimes$), Homare Murakami[2], Kien Nguyen[2], Kentaro Ishizu[2], Fumihide Kojima[2], Jong-Deok Kim[3], Sang-Hwa Chung[3], and Won-Joo Hwang[1]

[1] Department of Information and Communication System, Inje University, Gimhae 621-749, Gyeongnam, Korea
huynguyencse@gmail.com, ichwang@inje.ac.kr
[2] Wireless Systems Laboratory, Wireless Networks Research Center, National Institute of Information and Communications Technology, Yokosuka 239-0847, Japan
{homa,kienng,ishidu,f-kojima}@nict.go.jp
[3] School of Electrical and Computer Engineering, Pusan National University, Busan 46241, South Korea
{kimjd,shchung}@pusan.ac.kr

Abstract. Millimeter-Wave (mmWave) communication in ultra-dense networks (UDNs) has been considered as a promising technology for future wireless communication systems. Exploiting the benefits of mmWave and UDNs, we introduce a new approach for jointly optimizing small-cell base station (SBS) - user (UE) association and power allocation to maximize the system energy efficiency (EE) while guaranteeing the quality of service (QoS) constraints for each UE. The SBS-UE association problem poses a new challenge since it reflects as a complex mixed-integer non-convex problem. On the other hand, the power allocation problem is in non-convexity structure, which is impossible to handle with the association problem concurrently. An alternating descent method is thus introduced to divide the primal optimization problem into two sub-problems and handle one-by-one at each iteration, where the SBS-UE association problem is reformulated using the penalty approach. Then, path-following algorithms are developed to convert non-convex problem into the simple convex quadratic functions at each iteration. Numerical results are provided to demonstrate the convergence and low-complexity of our proposed schemes.

Keywords: Millimeter-Wave · mmWave · Ultra-dense networks · Energy efficiency · User association · Power allocation · Penalty · Successive convex programming

This research was supported by the MSIT (Ministry of Science and ICT), Korea, under the Grand Information Technology Research Center support program (IITP-2018-2016-0-00318) supervised by the IITP (Institute for Information & communications Technology Promotion).

T. Q. Duong et al. (Eds.): Qshine 2018, LNICST 272, pp. 87–101, 2019.
https://doi.org/10.1007/978-3-030-14413-5_7

1   Introduction

Ultra-dense networks (UDNs) by deploying multiple small-cells base stations (SBSs) such as micro-cells, pico-cells, micro-cell operators[1] distributed in the traditional cell have been emerged as a promising technology to improve the coverage and network performance [3]. In UDNs, the processing load can be effectively shared among SBSs instead of centralizing at the traditional macro base stations. However, due to the limited capacity and the scheduling scheme, such deployed SBSs are not able to serve a large number of users (UEs) simultaneously. Thus, the SBS-UE association problem becomes very important [15].

Due to the rapid shifting of high-speed broadband wireless access, the application of Gigabit services such as real-time streaming, gigabyte file transfer have been broadly developed [1,11]. The core of such services is mainly based on millimeter-Wave (mmWave) communication, which is one of the powerful technologies enabling the high data rate services [4,17]. Such mmWave technologies utilize the frequency band from 30 to 300 GHz, which corresponds to wavelengths from 10 to 1 mm [6]. It has been reported that the maximum attenuation of mmWave can achieve in the 60 GHz, 120 GHz, 180 GHz frequency bands [19].

The application of mmWave in UDNs has been widely studied to exploit the benefits of short-range communication to the high data-rate services [1,20]. In particular, the UE association schemes have been developed to improve the load balancing in the UDNs [18,20]. The resource allocation schemes were also investigated since the wireless channels in the mmWave network are strongly unstable due to high frequency operation [14]. In the extending mmWave investigation, the joint optimization for both UE association and power allocation has taken into account [20]. The binary association problem in those schemes is usually applied with the relaxation method, then using the traditional Lagrangian and gradient methods [20]. For those previous works, the problem will become very challenging when more UEs join in the system. Thus, the local optimum is difficult to achieve, which produces high-complexity with large iterations for convergence.

In this paper, a new approach for jointly optimizing the SBS-UE association and power allocation is investigated. The energy efficiency (EE) of the whole networks is taken into account subject to the quality of service (QoS) requirements of each UE. Since the proposed SBS-UE association problem is a challenging mixed integer non-convex optimization problem, which poses a high complexity for large-scale networks. On the other hand, the association problem and non-convex power allocation problem have a close connection. Therefore, solving them concurrently is extremely infeasible. Following from the proposed algorithms in [2], it has proved the effectiveness in the large-scale UDNs with low computational complexity. Particularly, an alternating descent method is first derived, which allows us to divide the primal problem into two separated

---

[1] The term "micro-cell operator" is used to indicate the deployment of micro-cell base stations at the private areas like school zones, factories, company buildings with their individual policies [5].

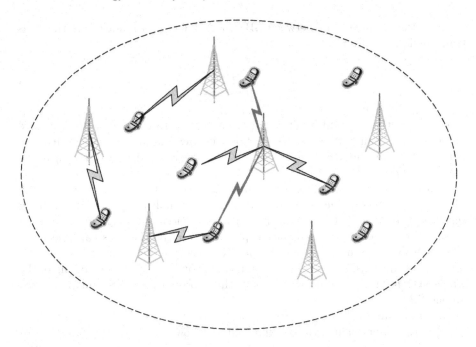

Fig. 1. The dense network scenario with M small-cell BSs and K users randomly distributed in the circular cell.

subproblems for easy to handle at the same time scale. Even though the association problem and power allocation problem are decoupled, their structures are still nonconvexity. The penalty method [15] is thus introduced, which relaxes those binary variables and force its square converting to its own original form. Then, the path-following procedures are employed to achieve computational low-complexity algorithms by converting the non-convex problem into the simple quadratic convex optimization problem [10,13,15]. Numerical results are thus developed to verify the convergence and effectiveness of our proposed algorithms.

Notation: Boldfaced symbols are used for optimization variables whereas non-boldfaced symbols are for deterministic terms, regardless of whether they are vectors or scalar values.

2   System Model and Problem Formulations

2.1   System Model

Consider the UDN consists of the set of M SBSs using 60-GHz mmWave band, $m \in \mathcal{M} \triangleq \{1, \ldots, M\}$ communicating with the set of K UEs, $k \in \mathcal{K} \triangleq \{1, \ldots, K\}$, where those locations are randomly distributed in the circular cell with radius $\mathcal{R}$, as shown in Fig. 1. All SBSs and UEs are equipped with single-antenna and operate in half-duplex mode. Since the 60-GHz frequency band is studied, the channel gain between SBS m and UE k is modeled as the Friis transmission equation [8]

$$g_{m,k} = \frac{l_{m,k}^{Tx} l_{m,k}^{Rx} \xi^2}{16\pi^2 (\frac{d_{mk}}{d_0})^\eta}, \tag{1}$$

where $l_{m,k}^{Tx}$ is the transmit antenna gain from SBS m to UE k, $l_{m,k}^{Rx}$ is the receive antenna gain from SBS m to UE k, ξ is the wavelength, d_{mk} is the distance between SBS m and UE k, d_0 is the far field reference distance, and η is the path-loss exponent ($\eta \in [2,6]$).

Without loss of generality, we assume that each UE can associate only one SBS at each interval time and each SBS can serve multiple UEs by performing scheduling algorithms to avoid intra-cell interferences [7]. There are two main reasons for proposing the association problem. Firstly, the UEs can select the nearest SBS or the associated SBS with the best channel gain such that their minimum transmission rates are guaranteed. Secondly, the load-balancing of the whole systems can be improved, where the processing load is equally shared among SBSs.

Let us define $\mathbf{x} \triangleq [\mathbf{x}_1, \ldots, \mathbf{x}_M]^T$, $\mathbf{x}_m \triangleq [\mathbf{x}_{m,1}, \ldots, \mathbf{x}_{m,K}]^T$, where $\mathbf{x}_{m,k} \in \{0,1\}$ is the association variable expressed as follow

$$\mathbf{x}_{m,k} = \begin{cases} 1 \text{ if UE } k \text{ associates with SBS } m, \\ 0 \text{ otherwise.} \end{cases}$$

Due to transmission scheduling among UEs at the associated SBS m, each UE will be assigned one time slot equal to another UEs. According to Shannon's capacity formula, the achievable rate for UE k via SBS m is given by

$$\mathcal{R}_{m,k}(\mathbf{x}, \mathbf{p}) = \frac{W}{\mathcal{K}_m(\mathbf{x})} \log_2(1 + \gamma_{m,k})$$

$$= \frac{W}{\mathcal{K}_m(\mathbf{x})} \log_2 \left(1 + \frac{\mathbf{p}_m g_{m,k}}{\sum\limits_{n \in \mathcal{M}\setminus\{m\}} \mathbf{p}_n g_{n,k} + \sigma_m^2} \right), \tag{2}$$

where W is the total bandwidth, $\mathbf{p}_m$ is the transmit power of SBS m, σ_m^2 is the variance of additive white Gaussian noise (AWGN), $\mathcal{K}_m(\mathbf{x}) = \sum_{k=1}^{K} \mathbf{x}_{m,k}$ is the number of UEs associated with SBS m. As seen from (2), UE k is subjected to the inter-cell interferences from other SBSs $n \neq m$.

2.2   Problem Formulations

Our objectives aim to maximize the EE maximization problem in terms of number of bits delivered per unit of Joule subject to the QoS rate threshold for each UE. The EE maximization problem can be expressed as [20].

$$\max_{\mathbf{x},\mathbf{p}} \mathcal{P}_1(\mathbf{x},\mathbf{p}) \triangleq \frac{\sum\limits_{m\in\mathcal{M}}\sum\limits_{k\in\mathcal{K}} \mathbf{x}_{m,k}\mathcal{R}_{m,k}(\mathbf{x},\mathbf{p})}{\sum\limits_{m\in\mathcal{M}} \mathbf{p}_m + p_{non}}, \tag{3a}$$

$$\text{s.t} \quad \sum_{m=1}^{M} \mathbf{x}_{m,k} = 1, \forall k \in \mathcal{K}, \tag{3b}$$

$$\mathbf{x}_{m,k} \in \{0,1\}, \forall m \in \mathcal{M}, k \in \mathcal{K}, \tag{3c}$$

$$0 \leq \mathbf{p}_m \leq \mathcal{P}_m^{max}, \forall m \in \mathcal{M}, k \in \mathcal{K}, \tag{3d}$$

$$\sum_{m\in\mathcal{M}} \mathbf{x}_{m,k}\mathcal{R}_{m,k}(\mathbf{x},\mathbf{p}) \geq \mathcal{R}_k^{min}, \forall k \in \mathcal{K}, \tag{3e}$$

where $p_{non} = M * p_a$ is the non-transmit power with p_a is the antenna circuit power. Constraint (3b) ensures that each UE must associate to one SBS while the constraint (3d) indicates the maximum transmit power of each SBS, and constraint (3e) guarantees the achievable rate of each UE must higher than the QoS requirement.

It can be observed from (3) that the objective function comprises of a difficult class of mixed-integer problem together with non-convex power allocation problem, which is a very challenging optimization problem. On the other hand, constraint (3e) also has nonconvexity structure. Therefore, dealing with binary association variable $\mathbf{x}$ and continuous transmit power variables $\mathbf{p}$ concurrently is infeasible. Especially at the large-scale networks, solving the association problem becomes more challenging. From the following, we introduce alternating descent algorithm, which allows us to divide the primal problem into two subproblems and handle one-by-one at the same time scale. In addition, the path-following methods with low-complexity are developed to convert non-convex problem into the simple quadratic convex problem at each iteration.

3 Alternating Descent Algorithm for Energy Efficiency Optimization Problem

In this section, we provide an approach to deal with those challenges in solving the problem (3). First of all, it can be observed that the binary constraint (3c) is a discrete variable. So finding the binary solution at the large scale networks takes very high-complexity and may increase in an exponential manner. Therefore, heuristic schemes such as binary search or exhaustive search seem infeasible to apply in this scenario. Tackling this issue, our aims are not only dealing with the binary in reasonable complexity but also focusing on high dimension networks. Following from [2,15], we first make the binary relaxation in constraint (3c) with box constraints as follows

$$\mathbf{x}_{m,k} \in [0,1], \forall m \in \mathcal{M}, k \in \mathcal{K}. \tag{4}$$

Realizing the characteristics of binary variables, we can easily observe that $x_{m,k}^2 \leq x_{m,k}, 0 \leq x_{m,k} \leq 1$. The equality holds true when $x_{m,k} \in \{0,1\}$. Since the binary variables and its square are in the box range, the equality only happens when they are approached 0 or 1. By exploiting those characteristics, we introduce the penalty approach in order to zero-force the subtraction between $x_{m,k}$ and its square. The new EE optimization problem can be reformulated as follows

$$\max_{x,p} \mathcal{P}_2(x,p) \triangleq \frac{\displaystyle\sum_{m \in \mathcal{M}} \sum_{k \in \mathcal{K}} x_{m,k}^2 \mathcal{R}_{m,k}(x^2,p)}{\displaystyle\sum_{m \in \mathcal{M}} P_m + p_{non}}$$

$$+ \Theta \sum_{m \in \mathcal{M}} \sum_{k \in \mathcal{K}} (x_{m,k}^2 - x_{m,k}),$$

$$\text{s.t} \quad (3b), (3d) - (3e), (4), \tag{5}$$

where Θ is the positive penalty factor which downgrade the gap of the subtraction term $(x_{m,k}^2 - x_{m,k})$ to zero.

Even though (5) is relaxed with continuous association variables, the optimization problem is still complex with non-convex structure. Therefore, finding the optimal solution for (5) still suffer from many difficulties due to the nonconcavity of the objective function and nonconvexity of the feasible sets. In the following, the alternating descent method is introduced to split the primal problem into two separated subproblems, which makes the problem easy to handle concurrently at the same iteration. In details, the SBS-UE association problem is handled while keeping the transmit power as a constant. Next, the transmit power is optimized based on the constant optimal association variables which found in the previous step. In such schemes, the increment of the objective function is guaranteed at each iteration until its convergence [15]. Since those subproblems are still in nonconvexity structures, the successive convex programming is developed to provide computationally low-complexity algorithms by solving the simple quadratic convex problem at each iteration [10,13,15].

3.1 SBS-UE Association Problem

By exploiting the alternating descent approach, we first focus on the SBS-UE association problem while ignoring the power allocation variables. Let us fix $p \triangleq p^{(\kappa)}$. Denote $t(p) = \left(\displaystyle\sum_{m \in \mathcal{M}} P_m + p_{non}\right)$. The optimization problem (5) remains as

$$\max_x \mathcal{P}_2(x,p) \triangleq \sum_{m \in \mathcal{M}} \sum_{k \in \mathcal{K}} x_{m,k}^2 \mathcal{R}_{m,k}(x^2,p)/t(p)$$

$$+ \Theta \sum_{m \in \mathcal{M}} \sum_{k \in \mathcal{K}} (x_{m,k}^2 - x_{m,k}),$$

$$\text{s.t} \quad (3b), (3e), (4). \tag{6}$$

By applying inequalities (22) for the UE achievable rate $(\mathbf{x}_{m,k}^2 \mathcal{R}_{m,k}(\mathbf{x}^2, p))$ and (24) for the penalty term $(\mathbf{x}_{m,k}^2 - \mathbf{x}_{m,k})$ in the Appendix to (6), we obtain the lower bound as

$$
\begin{aligned}
\mathcal{P}_2(\mathbf{x}, p) &\geq \sum_{m \in \mathcal{M}} \sum_{k \in \mathcal{K}} \lambda_{m,k}^{(\kappa)}(\mathbf{x}, p)/t(p) + \Theta \sum_{m \in \mathcal{M}} \sum_{k \in \mathcal{K}} \mu_{m,k}^{(\kappa)}(\mathbf{x}) \\
&\triangleq \mathcal{P}_2^{(\kappa)}(\mathbf{x}, p),
\end{aligned}
\tag{7}
$$

where

$$
\begin{aligned}
\lambda_{m,k}^{(\kappa)}(\mathbf{x}, p) &\triangleq \frac{(x_{m,k}^{(\kappa)})^2 \log_2(1 + \gamma_{m,k})}{\mathcal{K}_m((x_{m,k}^{(\kappa)})^2)} \\
&+ \frac{2 x_{m,k}^{(\kappa)} \log_2(1 + \gamma_{m,k})(\mathbf{x}_{m,k} - x_{m,k}^{(\kappa)})}{\mathcal{K}_m((x_{m,k}^{(\kappa)})^2)} \\
&- \frac{(x_{m,k}^{(\kappa)})^2 \log_2(1 + \gamma_{m,k})}{(\mathcal{K}_m((x_{m,k}^{(\kappa)})^2))^2} \\
&\quad (\mathcal{K}_m((\mathbf{x}_{m,k})^2) - \mathcal{K}_m((x_{m,k}^{(\kappa)})^2)), \\
\mu_{m,k}^{(\kappa)}(\mathbf{x}) &\triangleq ((x_{m,k}^{(\kappa)})^2 - x_{m,k}^{(\kappa)} + (2 x_{m,k}^{(\kappa)} - 1)(\mathbf{x}_{m,k} - x_{m,k}^{(\kappa)})).
\end{aligned}
\tag{8}
$$

The positive penalty factor Θ can be found at the initial step as follows

$$
\Theta = \left| \left(\sum_{m \in \mathcal{M}} \sum_{k \in \mathcal{K}} \lambda_{m,k}^{(0)}(\mathbf{x}, p)/t(p^{(0)}) \right) \backslash \left(\sum_{m \in \mathcal{M}} \sum_{k \in \mathcal{K}} \mu_{m,k}^{(0)} \right)(\mathbf{x}) \right|.
\tag{9}
$$

On the other hand, by applying inequality (23) in the Appendix for non-convex QoS constraint (3e) yields that

$$
\begin{aligned}
\sum_{m \in \mathcal{M}} \mathbf{x}_{m,k} \mathcal{R}_{m,k}(\mathbf{x}, p) &\geq \sum_{m \in \mathcal{M}} \log_2(1 + \gamma_{m,k}) \\
&\left(\frac{2\sqrt{x_{m,k}^{(\kappa)}} \sqrt{\mathbf{x}_{m,k}}}{\mathcal{K}_m(x_{m,k}^{(\kappa)})} - \frac{x_{m,k}^{(\kappa)}}{(\mathcal{K}_m(x_{m,k}^{(\kappa)}))^2} \mathcal{K}_m(\mathbf{x}_{m,k}) \right).
\end{aligned}
\tag{10}
$$

Thus, the SBS-UE association optimization problem can be expressed as follows

$$
\max_{\mathbf{x}} \mathcal{P}_2^{(\kappa)}(\mathbf{x}, p)
$$

$$
\text{s.t} \quad (3b), (4), (10).
\tag{11}
$$

Instead of finding the global optimum for (6), the problem can be targeted by solving lower bound maximization (11), which generates the feasible point $x^{(\kappa+1)}$

to improve the objective function in $x^{(\kappa)}$. Generally, at the initial point $x^{(0)}$, the optimization problem (11) will generate the set of sequence $x^{(\kappa)}, \kappa = 1, 2, \ldots$ such that

$$
\begin{aligned}
\mathcal{P}_2(x^{(\kappa+1)}, p) &\geq \mathcal{P}_2^{(\kappa)}(x^{(\kappa+1)}, p) \\
&\geq \mathcal{P}_2^{(\kappa)}(x^{(\kappa)}, p) \\
&= \mathcal{P}_2(x^{(\kappa)}, p).
\end{aligned}
\tag{12}
$$

In particular, $x^{(\kappa-1)}$ is used as a feasible point to obtain $x^{(\kappa)}$ until the convergence.

3.2 Power Allocation Problem

In the power allocation problem, we fix $\mathbf{x} \triangleq x^{(\kappa+1)}$, which is the optimal $\mathbf{x}^*$ found in the previous step. Thus, the power allocation problem is in the form

$$
\max_{\mathbf{p}} \mathcal{P}_2(x, \mathbf{p}) \triangleq \sum_{m \in \mathcal{M}} \sum_{k \in \mathcal{K}} x_{m,k}^2 \mathcal{R}_{m,k}(x^2, \mathbf{p})/t(\mathbf{p}),
$$

$$
\text{s.t} \quad (3d), (3e).
\tag{13}
$$

By applying inequality (21) in the Appendix to (5), we obtain the lower bound for $\mathcal{R}_{m,k}(x^2, \mathbf{p})$

$$
\mathcal{R}_{m,k}(x^2, \mathbf{p}) \geq
$$

$$
\alpha_{m,k} + \beta_{m,k} \left(1 - \frac{p_m^{(\kappa)}}{\mathbf{p}_m} + \frac{\displaystyle\sum_{n \in \mathcal{M} \backslash \{m\}} (p_n^{(\kappa)} - \mathbf{p}_n) g_{n,k}}{\displaystyle\sum_{n \in \mathcal{M} \backslash \{m\}} p_n^{(\kappa)} g_{n,k} + \sigma_m^2} \right),
$$

$$
\triangleq \mathcal{R}_{m,k}^{(\kappa)}(x^2, \mathbf{p}),
\tag{14}
$$

which is the concave function with

$$
0 \leq \alpha_{m,k} = \frac{W}{\mathcal{K}_m(x^2)} \log_2 \left(1 + \frac{p_m^{(\kappa)} g_{m,k}}{\displaystyle\sum_{n \in \mathcal{M} \backslash \{m\}} p_n^{(\kappa)} g_{n,k} + \sigma_m^2} \right),
$$

$$
0 \leq \beta_{m,k} = \frac{W}{\mathcal{K}_m(x^2)}
$$

$$
\left(\frac{(p_m^{(\kappa)} g_{m,k})/(\displaystyle\sum_{n \in \mathcal{M} \backslash \{m\}} p_n^{(\kappa)} g_{n,k} + \sigma_m^2)}{1 + (p_m^{(\kappa)} g_{m,k})/(\displaystyle\sum_{n \in \mathcal{M} \backslash \{m\}} p_n^{(\kappa)} g_{n,k} + \sigma_m^2)} \right).
\tag{15}
$$

Algorithm 1. Alternating descent algorithm for EE optimization algorithm

---

Initialization Initial any feasible point $x^{(0)}$, $p^{(0)}$. Run (18) to find a feasible point $p^{(\kappa)}$ for (5). Obtain Θ according to (9). Set $\kappa = 0$.
repeat
  Solve (11) with $\mathbf{p} = p^{(\kappa)}$ to find $x^{(\kappa+1)}$.
  Solve (16) with $\mathbf{x} = x^{(\kappa+1)}$ to find $p^{(\kappa+1)}$.
  Set $\kappa = \kappa + 1$.
until convergence

Thus, solving (13) is equal to find the feasible point $p^{(\kappa)}$ in the following problem

$$\max_{\mathbf{p}} \mathcal{P}_2^{(\kappa)}(x,\mathbf{p}) \triangleq \sum_{m\in\mathcal{M}}\sum_{k\in\mathcal{K}} x_{m,k}^2 \mathcal{R}_{m,k}^{(\kappa)}(x^2,\mathbf{p})$$

$$-\Xi(p^{(\kappa)})\mathbf{t}(\mathbf{p}), \tag{16a}$$

$$\text{s.t } (3d), \sum_{m\in\mathcal{M}} x_{m,k}\mathcal{R}_{m,k}^{(\kappa)}(x,\mathbf{p}) \geq \mathcal{R}_k^{min}, \forall k \in \mathcal{K}, \tag{16b}$$

where $\Xi(p^{(\kappa)})$ is obtained at each iteration as follows

$$\Xi(p^{(\kappa)}) = (\sum_{m\in\mathcal{M}}\sum_{k\in\mathcal{K}} x_{m,k}^2 \mathcal{R}_{m,k}(x^2,p^{(\kappa)}))\backslash(\mathbf{t}(p^{(\kappa)})). \tag{17}$$

Similar to the previous step, $p^{(\kappa)}, \kappa = 1,2,\ldots$ is found from the initial point $p^{(0)}$ by iteratively solving (16). Thus, the increment of the objective function is guaranteed since the optimal variables $p^{(\kappa+1)}$ improve the objective function at the $(\kappa+1)$-th iteration better than the objective function in the (κ)-th iteration.

3.3 Initialization

In order to solve (5), it is necessary to find the feasible point $\mathbf{p}$ for QoS constraint (3e). Let us define the initialization problem as follows

$$\max_{\mathbf{p}} \min_{k=1,\ldots,K} \sum_{m\in\mathcal{M}} \mathcal{R}_{m,k}^{(\kappa)}(x^2,\mathbf{p})$$

$$\text{s.t} \qquad (3d). \tag{18}$$

We run iteratively the problem (18) until the ratio of $(\sum_{m\in\mathcal{M}} \mathcal{R}_{m,k}^{(\kappa)}(x^2,\mathbf{p}))/(\mathcal{R}_k^{min})$ $\geq 1, \forall k \in \mathcal{K}$. Then reset $\kappa \Rightarrow 0$. It can be observe that the minimum achievable rate among UEs in (18) is increased at each iteration until the their QoS requirements are satisfied. The feasible initial point $\mathbf{p}$ is thus provided for solving (5).

From the initial point $x^{(0)}$, $p^{(0)}$, the above procedures improve the objective function in (5). In details, the SBS-UE association problem in (6) and the power

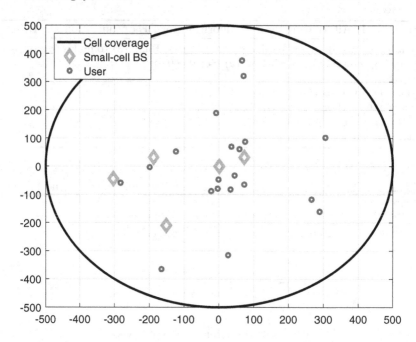

Fig. 2. The dense network scenario with $M = 5$ SBSs and $K = 20$ UEs randomly distributed in the circular cell.

allocation problem in (13) are alternatively solved to improve the problem (5), which make $(\mathbf{x}, \mathbf{p})$ converges to the optimal point at the finite iterations. The increment procedure for (5) can be expressed as

$$
\begin{aligned}
\mathcal{P}_2(\mathbf{x}^{(\kappa+1)}, \mathbf{p}^{(\kappa+1)}) &\geq \mathcal{P}_2^{(\kappa)}(\mathbf{x}^{(\kappa+1)}, \mathbf{p}^{(\kappa+1)}) \\
&\geq \mathcal{P}_2^{(\kappa)}(x^{(\kappa+1)}, \mathbf{p}^{(\kappa)}) = \mathcal{P}_2(\mathbf{x}^{(\kappa+1)}, \mathbf{p}^{(\kappa)}) \\
&\geq \mathcal{P}_2^{(\kappa)}(\mathbf{x}^{(\kappa+1)}, p^{(\kappa)}) \\
&\geq \mathcal{P}_2^{(\kappa)}(\mathbf{x}^{(\kappa)}, p^{(\kappa)}) = \mathcal{P}_2(\mathbf{x}^{(\kappa)}, \mathbf{p}^{(\kappa)}).
\end{aligned}
\tag{19}
$$

Note that, the convergence is activated with a small ϵ value when the below condition is triggered

$$
\left| \frac{\mathcal{P}_2(\mathbf{x}^{(\kappa+1)}, \mathbf{p}^{(\kappa+1)}) - \mathcal{P}_2(\mathbf{x}^{(\kappa)}, \mathbf{p}^{(\kappa)})}{\mathcal{P}_2(\mathbf{x}^{(\kappa)}, \mathbf{p}^{(\kappa)})} \right| \geq \epsilon.
\tag{20}
$$

The alternating descent method to solve EE maximization problem is summarized in Algorithm 1.

4   Numerical Results

In this section, Monte Carlo simulations are implemented to demonstrate the efficiency of our proposed algorithms. Consider the ultra-dense networks consisting of $M = 5$ SBSs and $K = 20$ UEs randomly distributed in the cell, as shown in Fig. 2. Following [20], we set $l_{m,k}^{Tx} = l_{m,k}^{Rx} = 1$, $\xi = 5$ mm, $\eta = 3$ and $d_0 = 1$ m. Without loss of generality, we assume that $\mathcal{P}_m^{max} = \mathcal{P}^{max}, \forall m \in \mathcal{M}$ and $\mathcal{R}_k^{min} = \mathcal{R}^{min}, \forall k \in \mathcal{K}$. The convergence threshold is set as $\epsilon = 1\mathrm{e}{-}4$. The other parameters are summarized by Table 1 [9, 20].

Table 1. Parameter settings

Parameter	Value
Cell radius	500 m
Bandwidth ($\mathcal{B}$)	1200 MHz
The maximum transmit power	4.7 dBm
Antenna power consumption	5.6 mW
Noise power density	−134 dBm/MHz

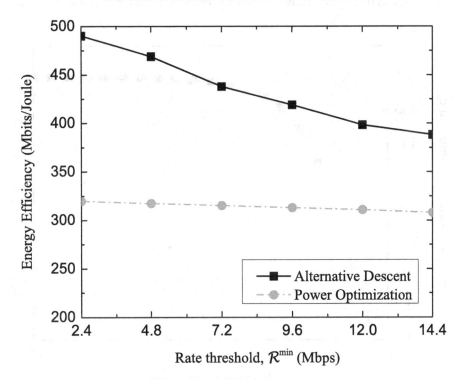

Fig. 3. The achievable EE versus $\mathcal{R}^{min}$ comparing to random SBS-UE association scenario.

In Fig. 3, we investigate the effects of UE QoS requirements on the achievable EE. We compare the achievable EE of the joint SBS-UE association and power allocation schemes, namely Alternative Descent with the random SBS-UE association scheme, namely Power Optimization. In the Power Allocation scheme, each UE selects the SBS which has the largest channel gain to associate, then the power allocation scheme is applied to optimize the transmit power in achieving EE maximization. From the plot, it can be observed that without SBS-UE association scheme, the achievable EE is strongly reduced while our proposed algorithm outperforms the Power Optimization scheme. On the other hand, we can observe from the figure that the Alternative Descent curve continuously decreases when $\mathcal{R}^{min}$ increases. This is due to the fact that when UE increases their QoS thresholds, all SBSs must increase their transmit power to meet the requirements while the total throughput is slightly scaled, which leads to the reduction on the achievable EE.

In Fig. 4, we demonstrate the computational low-complexity and convergence of our proposed algorithms. Observing from [20], solving those optimization schemes are usually based on the conventional Lagrangian and subgradient method. Therefore, their solutions are very complex and require large iterations for convergence. From the figure, it can be seen that our proposed solution is rapidly converged to the optimal performance, which requires around 12 itera-

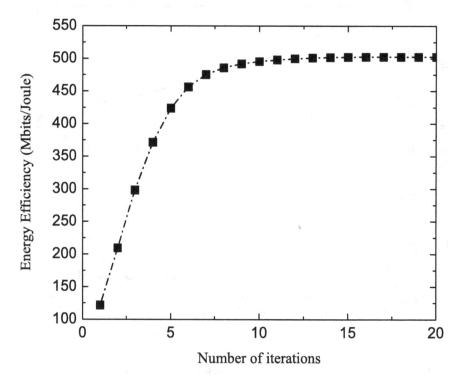

Fig. 4. Convergence of the proposed EE optimization with $r_{min} = 2.4$ Mbps.

tions. Thus, the results demonstrate the efficiency of our low-complexity solution, where the increment of the objective is guaranteed for each iteration.

5   Conclusions

In this paper, the new approaches for jointly optimizing SBS-UE association and power allocation have been proposed in the context of 60 GHz millimeter-wave ultra-dense networks. Our proposed methods have been targeted on maximizing the system EE subject to the QoS requirements of each UE. Since the SBS-UE association problem is one of the difficult classes of the mixed integer non-convex optimization problem, it is very challenging to solve together with non-convex power allocation scheme. Thanks to the help from alternating descent algorithm, the primal problem was divided into two suboptimization problems, which are handled one-by-one at each iteration. More specifically, the penalty approach has been applied to reformulate the challenging SBS-UE association problem. Finally, successive optimization methods were developed to reformulate non-convex optimization problem into low-complexity quadratic convex optimization problem with guaranteeing the increment of the objective at each iteration. The numerical results demonstrated the computational low-complexity and effectiveness of our proposed algorithms.

Appendix: Fundamental Inequalities

As the function $f(x, y) \triangleq \ln(1 + 1/xy)$ is convex in the domain $\{x > 0, y > 0\}$ [12], it follows that [16] for every $x > 0$, $y > 0$, $\bar{x} > 0$ and $\bar{y} > 0$,

$$
\begin{aligned}
\ln(1 + 1/xy) &= f(x, y) \\
&\geq f(\bar{x}, \bar{y}) + \langle \nabla f(\bar{x}, \bar{y}), (x, y) - (\bar{x}, \bar{y}) \rangle \\
&= \ln(1 + 1/\bar{x}\bar{y}) + \frac{1/\bar{x}\bar{y}}{1 + 1/\bar{x}\bar{y}}(2 - x/\bar{x} - y/\bar{y}).
\end{aligned}
\tag{21}
$$

Reutilizing inequalities in [15], we observe that function x^2/t is always convex under condition of $x > 0$ and $t > 0$, which yields inequality

$$
\frac{x^2}{t} \geq \frac{2\bar{x}}{\bar{t}}x - \frac{\bar{x}^2}{\bar{t}^2}t.
\tag{22}
$$

Then substituting $x \to \sqrt{x}$ and $\bar{x} \to \sqrt{\bar{x}}$, we obtain

$$
\frac{x}{t} \geq \frac{2\sqrt{\bar{x}}}{\bar{t}}\sqrt{x} - \frac{\bar{x}}{\bar{t}^2}t.
\tag{23}
$$

Lastly, the inequality

$$
x^2 - x \geq \bar{x}^2 - \bar{x} + (2\bar{x} - 1)(x - \bar{x})
\tag{24}
$$

always hold true since $x^2 - x$ is in convex quadratic form [15].

References

1. Bai, T., Heath, R.W.: Coverage and rate analysis for millimeter-wave cellular networks. IEEE Trans. Wirel. Commun. **14**(2), 1100–1114 (2015)
2. Che, E., Tuan, H.D., Nguyen, H.H.: Joint optimization of cooperative beamforming and relay assignment in multi-user wireless relay networks. IEEE Trans. Wirel. Commun. **13**(10), 5481–5495 (2014)
3. Chen, S., Qin, F., Hu, B., Li, X., Chen, Z.: User-centric ultra-dense networks for 5G: challenges, methodologies, and directions. IEEE Wirel. Commun. **23**(2), 78–85 (2016)
4. Gao, Z., Dai, L., Mi, D., Wang, Z., Imran, M.A., Shakir, M.Z.: MmWave massive-MIMO-based wireless backhaul for the 5G ultra-dense network. IEEE Wirel. Commun. **22**(5), 13–21 (2015)
5. Ishizu, K., Murakami, H., Ibuka, K., Kojima, F.: Next generation mobile communications system to realize flexible architecture and spectrum sharing. J. NICT **64**(2), 3–12 (2017)
6. Khan, F., Pi, Z.: mmWave mobile broadband (MMB): unleashing the 3–300GHz spectrum. In: 2011 34th IEEE Sarnoff Symposium, pp. 1–6, May 2011
7. Koivisto, M., Hakkarainen, A., Costa, M., Kela, P., Leppanen, K., Valkama, M.: High-efficiency device positioning and location-aware communications in dense 5G networks. IEEE Commun. Mag. **55**(8), 188–195 (2017)
8. Liu, P., Di Renzo, M., Springer, A.: Line-of-sight spatial modulation for indoor mmWave communication at 60 GHz. IEEE Trans. Wirel. Commun. **15**(11), 7373–7389 (2016)
9. Nguyen, L.D., Tuan, H.D., Duong, T.Q.: Energy-efficient signalling in QoS constrained heterogeneous networks. IEEE Access **4**, 7958–7966 (2016)
10. Nguyen, V.D., Duong, T.Q., Tuan, H.D., Shin, O.S., Poor, H.V.: Spectral and energy efficiencies in full-duplex wireless information and power transfer. IEEE Trans. Commun. **65**(5), 2220–2233 (2017)
11. Pi, Z., Choi, J., Heath, R.: Millimeter-wave gigabit broadband evolution toward 5G: fixed access and backhaul. IEEE Commun. Mag. **54**(4), 138–144 (2016)
12. Sheng, Z., Tuan, H.D., Nasir, A.A., Duong, T.Q., Poor, H.V.: Power allocation for energy efficiency and secrecy of wireless interference networks. IEEE Trans. Wirel. Commun. **PP**(99), 1–2 (2018)
13. Sheng, Z., Tuan, H.D., Duong, T.Q., Poor, H.V.: Joint power allocation and beamforming for energy-efficient two-way multi-relay communications. IEEE Trans. Wirel. Commun. **16**(10), 6660–6671 (2017)
14. Stephen, R.G., Zhang, R.: Joint millimeter-wave fronthaul and OFDMA resource allocation in ultra-dense CRAN. IEEE Trans. Commun. **65**(3), 1411–1423 (2017)
15. Tam, H.H.M., Tuan, H.D., Ngo, D.T., Duong, T.Q., Poor, H.V.: Joint load balancing and interference management for small-cell heterogeneous networks with limited backhaul capacity. IEEE Trans. Wirel. Commun. **16**(2), 872–884 (2017)
16. Tuy, H.: Convex Analysis and Global Optimization. Springer, Heidelberg (2017). https://doi.org/10.1007/978-3-319-31484-6
17. Wang, P., Li, Y., Song, L., Vucetic, B.: Multi-gigabit millimeter wave wireless communications for 5G: From fixed access to cellular networks. IEEE Commun. Mag. **53**(1), 168–178 (2015)
18. Xu, Y., Shokri-Ghadikolaei, H., Fischione, C.: Distrubuted association and relaying with fairness in millimeter wave networks. IEEE Trans. Wirel. Commun. **5**(12), 7955–7970 (2016)

19. Yang, G., Du, J., Xiao, M.: Maximum throughput path selection with random blockage for indoor 60 GHz relay networks. IEEE Trans. Commun. **63**(10), 3511–3524 (2015)
20. Zhang, H., Huang, S., Jiang, C., Long, K., Leung, V.C., Poor, H.V.: Energy efficient user association and power allocation in millimeter-wave-based ultra dense networks with energy harvesting base stations. IEEE J. Sel. Areas Commun. **35**(9), 1936–1947 (2017)

Priority-Based Device Discovery in Public Safety D2D Networks with Full Duplexing

Zeeshan Kaleem[1(✉)], Muhammad Yousaf[2], Syed Ali Hassan[3], Nguyen-Son Vo[4], and Trung Q. Duong[5]

[1] Electrical Engineering Department,
COMSATS University Islamabad, Wah Campus, Wah Cantonment, Pakistan
zeeshankaleem@gmail.com
[2] Telecommunications Engineering Department, UET Taxila, Taxila, Pakistan
myousafalamgir@yahoo.com
[3] School of Electrical Engineering and Computer Science (SEECS),
National University of Sciences and Technology (NUST), Islamabad, Pakistan
ali.hassan@seecs.edu.pk
[4] Duy Tan University, Da Nang, Vietnam
vonguyenson@gmail.com
[5] School of Electronics, Electrical Engineering and Computer Science,
Queen's University Belfast, Belfast, UK
trung.q.duong@qub.ac.uk

Abstract. Device-to-device (D2D) services are gaining popularity in public safety (PS) applications. The existing half-duplex (HD) D2D discovery has the constraint that the devices sending beacon cannot be discovered at the same time, resulting in large time delays. To counter this problem, in-band full duplex (IB-FD) communications can be used to discover the user quickly by enabling simultaneous transmission and reception during same time-frequency block. In this paper, we exploit the benefits of IB-FD system where PS users are given priority in resource allocation. Moreover, to efficiently utilize the spectrum, we propose a time-efficient device discovery resource allocation (TE-DDRA) scheme where a user can switch the transmission mode from HD to IB-FD when the demand exceeds the available resources in HD mode. The simulation results prove that in comparison with random mode, the PS priority mode saves around 37% discovery time.

Keywords: D2D discovery · Full duplex device discovery ·
5G systems · Public safety

This work was supported by the Newton Prize 2017 and a Research Environment Links grant, ID 339568416, under the Newton Programme Vietnam partnership. The grant is funded by the UK Department of Business, Energy and Industrial Strategy (BEIS) and delivered by the British Council. For further information, please visit www.newtonfund.ac.uk./.

T. Q. Duong et al. (Eds.): Qshine 2018, LNICST 272, pp. 102–108, 2019.
https://doi.org/10.1007/978-3-030-14413-5_8

1 Introduction

The fifth-generation (5G) systems are targetting high data rate and low-latency demands of the mobile users [2,3,11,14,15]. To meet these demands, the deployment of device-to-device (D2D) communications [17] attracts the service providers because of their capability to provide efficient spectrum utilization. D2D communications can allow data flow from one device to another without the help of a base station (BS) resulting in increased data rate and reduced latency [5]. However, prior to link establishment, the discovery of devices lying in the neighborhood is the first and important phase. The discovery should be completed in a short time to maintain the 5G latency requirements by utilizing the limited available spectrum.

D2D discovery underlaying cellular architecture can be performed under different levels of operator controls. The third generation partnership project (3GPP) categorized the D2D discovery as fully controlled (user-specific type-2) and partially controlled (user-specific type-1) [1]. We implement user-specific type-2 discovery scheme for public safety (PS) users, whereas for non-PS (NPS) users we consider the user-specific type-1 discovery [9]. Since NPS users have no strict requirements, therefore we implement user-specific type-1 discovery scheme for them (Fig. 1).

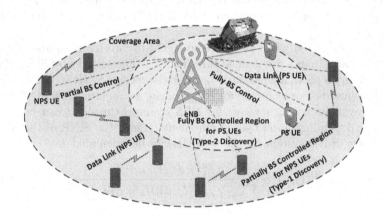

Fig. 1. Types of D2D discovery schemes for public safety networks.

In literature, numerous power control [8,16] and scheduling schemes [10,12] has been implemented to reduce interference among the heterogeneous networks, but unfortunately these schemes cannot be implemented for D2D scenario. Moreover, some solutions related to D2D discovery resource allocation such as location-based device discovery [7], device discovery to enable proximity services and [4] also exists in the literature. These schemes reduce the discovery time of users only for half-duplex (HD) mode. Therefore, their discovery time inevitably increases as compared to in-band full-duplex (IB-FD) mode.

In IB-FD system, a device can transmit and receive the beacon during the same time frequency block, which helps to reduce the discovery time. Previously, due to excessive self-interference (SI), IB-FD systems were considered impractical. However, recent advancements in SI cancellation make it feasible and practical for implementation in D2D discovery [6].

In this paper, we propose a time-efficient device discovery resource allocation (TE-DDRA) scheme for IB-FD system. The proposed TE-DDRA scheme has the option to switch from HD mode to IB-FD mode when the demand of resource allocation is not met in HD mode. Moreover, to reduce the discovery time for PS users, this scheme uses multi-channel ALOHA (MC-ALOHA) with energy sensing (MCALOHA-ES)[13] to reduce discovery time. The NPS user on the other hand, use conventional ALOHA.

2   System Model

We consider a multi-cell deployment scenario with a single cell as an example. In the multi-cell system, each cell has a radius R and an enhanced node B (eNB) at the origin. Moreover, a total of K users are deployed in the coverage area that follow a homogeneous Poisson point process (PPP) with density λ. We assume N PS user equipments (PS UEs) and M NPS UEs. Furthermore, we consider that every k-th user is transmitting a beacon with maximum power $P_k^{\max}$. Also, the uplink (UL) Long-term evolution advanced (LTE-A) system with bandwidth B is considered.

A UE is discovered when the received signal-to-interference-plus-noise-ratio (SINR) is greater than a pre-defined threshold γ^{th}. The selection of SINR threshold depends upon particular scenario because the receive SINR is different for PS and NPS scenarios. For example, for PS scenario, we require a low SINR threshold to discover far users, however for NPS scenario threshold can be selected high or low depending upon QoS requirements. Thus, the SINR of the received signal in case of HD system for the 0-th UE can be calculated as

$$\gamma_0^{\text{HD}}(j,d) = \frac{P_s^t(j,d)|H_s(j,d)|^2}{\sum_{k=1,k\neq s}^{K} P_k^t(j,d)|H_k(j,d)|^2 + \sigma^2}, \tag{1}$$

where $|H_s(j,d)|^2$ is the squared envelope of the channel gain between the transmitting UE and the 0-th receive UE on resource block d during the j-th subframe, and σ^2 is the noise variance. In HD system, a UEs transmitting the beacon cannot listen to beacon at the same time which results in less number of discovered users and increased interference. In this paper, the PS user can switch from HD and IB-FD mode if the available resources are less than the demanded resources.

Similarly, the SINR of the received signal in case of IB-FD system for the 0-th PS UE can be calculated as

$$\gamma_0^{\text{FD}}(j,d) = \frac{P_s^t(j,d)|H_s(j,d)|^2}{\overline{P_0^t(j,d)}\overline{|H_0(j,d)|}^2 + \sum_{k=1,k\neq s}^{K-M} P_k^t(j,d)|H_k(j,d)|^2 + \sigma^2}. \tag{2}$$

where $\overline{H_0} = \beta H_0$, $\overline{x_0}$ and H_0 is the transmitting data causing SI and channel coefficient from 0-th user transmit antenna to its own receive antenna, respectively. The factor $\beta (0 \leq \beta \leq 1)$ is self-interference cancellation factor at the 0-th UE where $\beta = 0$ indicates perfect SI cancellation.

3 Proposed Time Efficient Discovery Resource Allocation Scheme

The discovery of more users in less time can be achieved by decreasing the number of collisions among the users transmitting beacons simultaneously. In the proposed TE-DDRA, the available discovery resources are split into two portion: dedicated resources for PS users and a pool of resources for NPS users, thereby completely avoiding contention among two categories. The contention can be further reduced for PS users by using multi-channel ALOHA with energy sensing (MCALOHA-ES) after a specified time frame. The MCALOHA-ES scheme senses the collision first before transmitting the beacon which reduces the chances of collisions. The MCALOHA-ES selects the discovery resource block where the received energy on d-th resource block is minimum, implying that no other user is transmitting the beacon on that RB. We summarize the major steps of the proposed TE-DDRA schemes in Algorithm 1.

4 Simulation Results

Simulation Environment: System-level simulations are performed to check the validity of the proposed TE-DDRA scheme. The key simulation parameters are summarized in Table 1.

Simulation Scenarios: We consider two main scenarios; (1) conventional random HD/FD scenario, and (2) proposed TE-DDRA HD/FD scenario. In conventional random HD/FD scenario, we consider the existing D2D discovery frame structure and treat all users equally, in sense of accessing D2D discovery resource blocks. Moreover, HD/FD users only use MCALHOHA scheme to access the discovery resources. Contrary to that, the proposed TE-DDRA HD/FD scheme uses the IB-FD frame structure with dedicated bands for PS users during the discovery phase to reduce the contention among users. Furthermore, this scheme has capability to switch the channel access protocol from MCALOHA to MCALOHA-ES after a predefined time duration.

Discovery time plays a critical role especially for PS situation. We evaluate the discovery time versus the number of users discovered conventional HD/FD mode without prioritizing PS users, and the proposed TE-DDRA with IB-FD mode and prioritization to PS users. In Fig. 2, we compare the performance of conventional random HD/FD with the proposed TE-DDRA scheme. This gain is obtained in IB-FD mode because there is no constraint of receiving the discovery beacon during the beacon transmitting period. Hence, more users can

---

Algorithm 1. Proposed TE-DDRA Scheme

---

1 **Initialization:** $T = 1$, U $= \{1, 2, ..., K\}$, M PS users, N NPS users,
   $\gamma^{\text{th}} = 4.5$ dB, $P^{\text{max}} = 23$ dBm, $j = 1$

2 **Assumptions:** All users are initially connected in HD mode, and will always
   transmit with power P^{max}

3 **Discovery resource request**: During current discovery period T, K users
   send resource request to eNB

4 **while** *K users discovered* **do**

5   **if** *No. of RB $\leq$ K user sending request for RB* **then**

6   Switch M PS users to IB-FD mode

7   Continue using HD mode for N NPS users

8   **if** *j $\leq$ W* **then**

9   Allocate resources among M PS users and N NPS users using
          MCALOHA

10   **else**

11   Allocate resources among N NPS users using MCALOHA

12   Use MCALOHA-ES for M PS users resource allocation

13   **end**

14   **if** $\gamma_k \geq \gamma^{\text{th}}$ **then**

15   Neighbor user are successfully discovered by the user sending
          beacon

16   **else**

17   Undiscovered user will wait for upcoming j-th subframe to receive
          the beacon

18   **end**

19        $j = j + 1$

20   **else**

21   Continue as HD mode for K users

22   Allocate resources among K users using MCALOHA

23   **if** $\gamma_k \geq \gamma^{\text{th}}$ **then**

24   Neighbor user are successfully discovered by the user sending
          beacon

25   **else**

26   Undiscovered user will wait for upcoming j-th subframe to receive
          the beacon

27   **end**

28        $j = j + 1$

29   **end**

30     $T = T + 1$

31 **end**

---

be discovered within less time frame. It can be seen that the proposed method discover more number of users at varying at varying SI cancellation values as shown in Fig. 2.

Table 1. Simulation parameters.

Parameters	Values
LTE layout	1 site (3 cells)
System bandwidth, RBs	10 MHz, 50 RBs
Carrier frequency	2 GHz
UE parameters	max Tx power: 23 dBm, noise figure: 9 dB
Threshold values	$\gamma^{\text{th}} = 4.5\,\text{dB}$, $W = 3$
Discovery resource selection schedulers	PS user: MCALOHA or MCALOHA-ES NPS user: MCALOHA
PL model	WINNER + B1
Fast fading	PedB

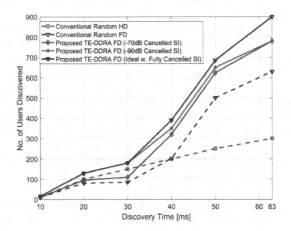

Fig. 2. Discovery time vs. number of discovered users.

5 Conclusion

In this paper, concept of switching the duplex mode from half-duplex to IB-FD is implemented that considerably reduces the discovery time for PS users. From simulation results, we proved the validity of the proposed scheme as it saves around 37% of discovery time as compared to conventional random access system.

References

1. 3GPP: Study on LTE device to device proximity services - Radio aspects. Technical report, 3GPP TR 36.843 v 12.0.1 (2014)
2. Ahmad, I., Kaleem, Z., Narmeen, R., Nguyen, L.D., Ha, D.B.: Quality-of-service aware game theory-based uplink power control for 5G heterogeneous networks. Mob. Netw. Appl. 1–8 (2018)

3. Andrews, J.G., et al.: What will 5G be? IEEE J. Sel. Areas Commun. **32**(6), 1065–1082 (2014)
4. Choi, K.W., Han, Z.: Device-to-device discovery for proximity-based service in LTE-advanced system. IEEE J. Sel. Areas Commun. **33**(1), 55–66 (2015)
5. Feng, D., Lu, L., Yuan-Wu, Y., Li, G.Y., Li, S., Feng, G.: Device-to-device communications in cellular networks. IEEE Commun. Mag. **52**(4), 49–55 (2014)
6. Heino, M., et al.: Recent advances in antenna design and interference cancellation algorithms for in-band full duplex relays. IEEE Commun. Mag. **53**(5), 91–101 (2015)
7. Hu, L.: Resource allocation for network-assisted device-to-device discovery. In: International Conference on Wireless Communications, Vehicular Technology, Information Theory and Aerospace & Electronic Systems (VITAE), pp. 1–5. IEEE (2014)
8. Kaleem, Z., Ahmad, A., Rehmani, M.H.: Neighbors' interference situation-aware power control scheme for dense 5G mobile communication system. Telecommun. Syst. **67**, 1–8 (2017)
9. Kaleem, Z., Chang, K.: Public safety priority-based user association for load balancing and interference reduction in PS-LTE systems. IEEE Access **4**, 9775–9785 (2016)
10. Kaleem, Z., Chang, K.: QoS priority-based coordinated scheduling and hybrid spectrum access for femtocells in dense cooperative 5G cellular networks. Trans. Emerg. Telecommun. Technol. **29**(1), e3207 (2018)
11. Kaleem, Z., Hui, B., Chang, K.: QoS priority-based dynamic frequency band allocation algorithm for load balancing and interference avoidance in 3GPP LTE HetNet. EURASIP J. Wirel. Commun. Netw. **2014**(1), 185 (2014)
12. Kaleem, Z., Khaliq, M.Z., Khan, A., Ahmad, I., Duong, T.Q.: PS-CARA: context-aware resource allocation scheme for mobile public safety networks. Sensors **18**(5), 1473 (2018)
13. Kaleem, Z., Li, Y., Chang, K.: Public safety users' priority-based energy and time-efficient device discovery scheme with contention resolution for prose in third generation partnership project long-term evolution-advanced systems. IET Commun. **10**(15), 1873–1883 (2016)
14. Kim, W., Kaleem, Z., Chang, K.: Power headroom report-based uplink power control in 3GPP LTE-A HetNet. EURASIP J. Wirel. Com. Networking **2015**(1), 233 (2015)
15. Kim, W., Kaleem, Z., Chang, K.: UE-specific interference-aware open-loop power control in 3GPP LTE-A uplink HetNet. In: International Conference on Ubiquitous and Future Networks (ICUFN), pp. 682–684. IEEE (2015)
16. Kim, W., Kaleem, Z., Chang, K.: Interference-aware uplink power control in 3GPP LTE-A HetNet. Wirel. Pers. Commun. **94**, 1–15 (2017)
17. Li, Y., Kaleem, Z., Chang, K.: Interference-aware resource-sharing scheme for multiple D2D group communications underlaying cellular networks. Wirel. Personal Commun. **90**(2), 749–768 (2016)

Modified Direct Method
for Point-to-Point Blocking Probability
in Multi-service Switching Networks
with Resource Allocation Control

Mariusz Głąbowski$^{(\boxtimes)}$ (iD), Maciej Sobieraj, and Maciej Stasiak

Faculty of Electronics and Telecommunications, Poznan University of Technology,
Poznań, Poland
mariusz.glabowski@put.poznan.pl

Abstract. This article proposes a simplified approach to the internal blocking probability calculation in switching networks with mechanisms controlling resource allocation for offered multi-service traffic streams. This resource allocation control can be executed using the so-called threshold and resource reservation mechanisms, according to which the volume of resources admitted depends on a traffic class and on the occupancy state of the interstage and outgoing links of the switching network. The developed method is of generic nature and allows one to model switching systems regardless of the implemented resource allocation control mechanism. However, despite its generic character, the method provides better accuracy as compared to the methods worked out earlier.

Keywords: Switching networks · Teletraffic ·
Performance evaluation · Elastic optical networks

1 Introduction

One of the basic elements that influences the efficiency and effectiveness of telecommunications networks is, besides their structure and traffic routing rules, network nodes and network devices like routers and switches that are used to connect services in these networks. All high-performance network devices contain a switching structure, implemented in the form of a switching network. Available studies and investigations [2,12,13] clearly show that the multi-stage Clos switching network is one of the most effective structures that provide high scalability. The universality of the Clos switching network also allows it to be applied to large data centres and elastic optical networks [10]. Multi-stage switching networks, both electronic and optical, even though they have a number of substantial differences, share many similarities, which makes it possible to model them at the level of streams (flows, calls) using the so-called *multi-rate* models [1,9].

© ICST Institute for Computer Sciences, Social Informatics and Telecommunications Engineering 2019
Published by Springer Nature Switzerland AG 2019. All Rights Reserved
T. Q. Duong et al. (Eds.): Qshine 2018, LNICST 272, pp. 109–118, 2019.
https://doi.org/10.1007/978-3-030-14413-5_9

Currently, among a large number of methods for modeling switching networks, the so-called *effective availability methods* are decidedly dominant. According to the effective-availability concept, blocking probability in a multi-stage switching network with multi-rate (multi-service) traffic are determined using equivalent single-stage single-service model, i.e., Ideal Erlang's Grading [3,11]. Initially, these methods were used to model switching networks that did not differentiate the quality of service for individual call streams [11]. The analytical studies that made it possible to take into account appropriate resource reservation algorithms for individual traffic streams in inter-stage links and outgoing links of a switching network are proposed, among others, in [8]. The works that followed deal with switching networks with elastic and adaptive services (implemented using threshold mechanisms) have been proposed in [7]. A generalized recurrent method for analytical calculation of blocking probability in multi-stage switching networks with implemented mechanisms for controlling resource allocation is proposed in [4]. The unification of the modeling patterns for different mechanisms for controlling resource allocation for particular traffic streams made it then possible, in the process of a determination of blocking probability in switching networks, to take into consideration traffic streams that are generated according to the following distributions: Erlang, Engset and Pascal. The recurrent method for modelling switching networks used in [4] is dedicated to multi-stage networks with a greater number of stages (greater than 3). This method was initially applied to networks without resource allocation control [5]. A modification to the recurrent method, introduced in [4], that provided better accuracy of calculations of the internal blocking, can be also applied in the case of the remaining effective availability methods that allow the point-to-point blocking probability to be determined, i.e. the PPBMT method and the direct method [8] that are dedicated to three-stage networks.

The present article discusses the modified direct method for determining point-to-point blocking probability in switching networks with introduced mechanisms for controlling resource allocation for multi-service traffic flows. The remaining part of the article is divided as follows. In Sect. 2 a basic model of the switching network is presented. Section 3 presents the method for determining the external blocking probability and the modified method for a determination of the internal blocking probability. Section 4 includes the numerical results that allow the accuracy of the introduced modifications to be evaluated. Section 5 sums up the most important results presented in the paper.

2   General Traffic Model of the Switching Network

Let us consider the multi-service ($z = 3$)-stage blocking switching networks [4] with the Clos structure (Fig. 1). Each of z stages is composed of v square switches with $v \times v$ links. The capacity of inter-stage links is limited to f allocation units (AUs). The output links of the switching network are grouped in the so-called outgoing directions that lead to the neighboring nodes. In the case of a typical execution of output directions, the direction r is composed of v links outgoing from the switches of the z stage.

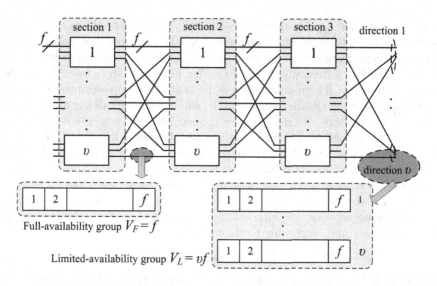

Fig. 1. 3-stage switching network

This network is offered traffic streams generated by three types of traffic sources, defined in traffic theory, i.e. Erlang, Engset and Pascal sources. Each of these types is related to different call streams, described by the flow intensity (call arrival) and the number of required AUs. The network under consideration is offered m_I Erlang traffic streams, m_J Engset traffic streams and m_K Pascal traffic streams. A call of any class c ($1 \leq c \leq m$, where $m = m_I + m_J + m_K$) requires t_c AUs for a service. Service times for calls of class c follows an exponential distribution with the parameter μ_c. Traffic intensity A_i of class i ($1 \leq i \leq m_I$) Erlang stream, traffic intensity $A_j(n)$ of class j ($1 \leq j \leq m_J$) Engset stream and traffic intensity of class k ($1 \leq k \leq m_K$) Pascal stream $A_k(n)$ are equal:

$$A_i = \lambda_i/\mu_i, \tag{1}$$
$$A_j(n) = (S_j - n_j(n))\Lambda_j/\mu_j, \tag{2}$$
$$A_k(n) = (S_k + n_k(n))\gamma_k/\mu_k, \tag{3}$$

where: n – the number of currently occupied allocation units – network state; λ_i – call arrival intensity for Erlang calls (according to Poisson distribution); S_j – the cardinality of Engset sources of class j; $n_j(n)$ – the number of serviced calls of class j in the state of n AU being busy; Λ_j – call arrival intensity for a free Engset source of class j; S_k – the cardinality of Pascal sources of class k; $n_k(n)$ – the number of serviced calls of class k in the state of n AU being busy; γ_k – call arrival intensity for a free Pascal source of class k.

Switching networks can operate in different modes for output link selection of links that lead to a successive node over a connection path. One of the most frequently used selection modes considered in traffic engineering is the point-to-point selection and point-to-group selection. According to the point-to-point selection, which is the subject of the present article, the process of finding a path for a new call (stream) of class c starts with a determination of a switch of the last stage that has a free outgoing link (having not less than t_c free AUs) in the demanded direction. Then, an attempt is made at setting up a connection inside the network, between the first stage switch (where a class c call has arrived) and the selected last stage switch. If a free connection path is available, then the connection is successfully executed. Otherwise, the call is lost (blocked) due to the occurrence of internal blocking phenomenon. The call can be also lost due to the occurrence of external blocking phenomenon, i.e., when there is no last stage switch having at least one outgoing link having no less than t_c AUs. In order to calculate the total blocking probability $E_{\mathrm{Tot}}(c)$ for class c calls offered to the switching network, the internal and external blocking phenomena should be taken into account, according to the following formula:

$$E_{\mathrm{Tot}}(c) = E_{\mathrm{Ex}}(c) + E_{\mathrm{In}}(c)\left[1 - E_{\mathrm{Ex}}(c)\right], \tag{4}$$

where $E_{\mathrm{Ex}}(c)$ denotes the external blocking probability, while $E_{\mathrm{In}}(c)$ denotes the internal blocking. The form of the Formula (4) results from the operation of the algorithm for the selection of a connection path in the switching network with the point-to-point selection.

3   Total Blocking Probability

3.1   External Blocking Probability

The blocking probability in outgoing directions is usually determined on the basis of a model of the limited-availability group (LAG). This probability depends on the structure of LAG (the number of outgoing links, the capacity of outgoing links) and on the introduced mechanisms to control the volume of allocated resources, i.e. on threshold mechanisms and reservation mechanisms.

The Structure of LAG. The LAG model is a model of the distributed system composed of v separated links having the capacity of f AU. The total capacity of the outgoing direction (multi-service system) $V_L = vf$ AU. A new call can be admitted for service only when it can be serviced by the resources of one of available links. The influence of the structure on traffic characteristics is taken into consideration in analytical models by an introduction of the conditional transition coefficients $\sigma_{c,S_L}(n)$:

$$\sigma_{c,S_L}(n) = [F(V_L - n, v, f, 0) - F(V_L - n, v, t_c - 1, 0)]/F(V_L - n, v, f, 0), \tag{5}$$

where $F(x, v, f, t)$ allows us to calculate the number of possible arrangements of x of free AUs in v links, each of which having the capacity equal to f AUs, with an additional assumption that there are at least t free (unoccupied) AUs in each link:

$$F\left(x, v, f, t\right) = \sum_{r=0}^{\left\lfloor \frac{x - vt}{f - t + 1} \right\rfloor}(-1)^r\binom{v}{r}\binom{x - v(t-1) - 1 - r(f - t + 1)}{v - 1}. \tag{6}$$

Threshold Mechanism in Output Directions. Threshold mechanisms make it possible to dynamically change the amount of resources admitted to particular service classes on the basis of the number of AUs being occupied in the outgoing directions of a switching network [7]. When a certain predefined limit of the occupancy of the outgoing direction in a switching network is exceeded, then a decrease in the resources allocated to calls of a given service class follows. As a result, this leads to an extension of the service time (for elastic services, in which data have to be transmitted in their entity), or to a decreasing of bitrate of transmitted multi-media data (for adaptive services, in which a prolongation of service time is not allowed) [7].

In analytical models of threshold systems it is assumed that each class c has a defined set of thresholds: $\{Q_{c,1}, Q_{c,2}, ..., Q_{c,q_c}\}$, where $Q_{c,1} \leq Q_{c,2} \leq ... \leq Q_{c,q_c}$. The further assumption is that the threshold area u of class c, limited by the threshold $Q_{c,u}$ and threshold $Q_{c,u+1}$, is defined by the given set of parameters $\{t_{c,u}, \mu_{c,u}\}$, where $t_{c,0} > t_{c,1} > ... > t_{c,u} > ...t_{c,q_c}$ and $\mu_{c,0}^{-1} \leq \mu_{c,1}^{-1} \leq ... \leq \mu_{c,u}^{-1} \leq ... \leq \mu_{c,q_c}^{-1}$. The parameter $\sigma_{c,\mathrm{T},u}(n)$ determines the occupancy states of threshold area u in which offered traffic is defined by the given parameters $t_{c,u}$ and $\mu_{c,u}$:

$$\sigma_{c,\mathrm{T},u}(n) = \begin{cases} 1 & \text{for } Q_{c,u} < n \leq Q_{c,u+1}, \\ 0 & \text{for remaining } n. \end{cases} \tag{7}$$

Let us notice that after applying threshold mechanisms in outgoing links (modeled by the LAG) we have introduced another type of dependency (besides the one imposed by the group structure) to the service process occurring in the switching networks under consideration. Since the threshold mechanism does not depend on the structure of the outgoing direction (only on its occupancy), it provides an opportunity for a product-form determination of the transition coefficient in LAG $\sigma_{c,S,u}(n)$:

$$\sigma_{c,S,u}(n) = \sigma_{c,S_L}(n) \cdot \sigma_{c,\mathrm{T},u}(n). \tag{8}$$

By taking into consideration the conditional transition coefficient it is possible to determine the occupancy state for each area u [7]:

$$\begin{aligned} n\left[Q_n\right]_{V_L} = & \sum_{i=1}^{m_I} \sum_{u=0}^{q_i} A_i t_i \sigma_{i,S,u}(n - t_i)\left[Q_{n-t_i}\right]_{V_L} \\ & + \sum_{j=1}^{m_J} \sum_{u=0}^{q_j} A_j(n)\sigma_{j,S,u}(n - t_j)t_j\left[Q_{n-t_j}\right]_{V_L} \\ & + \sum_{k=1}^{m_K} \sum_{u=0}^{q_k} A_k(n)\sigma_{k,S,u}(n - t_k)t_k\left[Q_{n-t_k}\right]_{V_L}. \end{aligned} \tag{9}$$

Bandwidth Reservation Mechanism in Output Directions. The reservation mechanism can be treated as a particular case of a threshold mechanism. According to the reservation mechanism, calls of particular classes cannot be admitted for service in a certain space of occupancy states of groups. The reservation mechanism is usually used to enforce an appropriate access to the resources of a system for all traffic classes. In a particular case, the reservation mechanism ensure equal blocking probability for all traffic classes, regardless their resource requirements. In analytical models of systems with reservation, the so-called *reservation limit/boundary* R_c for each traffic class is defined. The parameter R_c defines such a particular state of the system in which it can still admit a class c call. Note that an analysis of thus considered system can be brought in its essence to an analysis of a threshold system in which, after the last threshold is exceeded, the volume of allocated resources is equal to zero.

Determination of the External Blocking Probability. The external blocking probability $E_{\mathrm{Ex}}(c)$ can be calculated using LAG model of the outgoing directions in the switching network [7]. To determine the external blocking probability it is necessary to first calculate the occupancy distribution in LAG using the generalized Kaufman-Roberts distribution (9). After the occupancy distribution $[Q_n]_V$ has been determined, the blocking probability of calls of particular traffic classes in LAG, regardless of the applied resource allocation control mechanism, can be expressed by the following formula:

$$E_{\mathrm{Ex}}(c) = \sum_{n=V_L-v(t_{c,q_c}-1)}^{V_L} [Q_n]_{V_L}. \tag{10}$$

3.2   Internal Blocking Probability

Determination of the Effective Availability Parameter. Internal blocking phenomenon decreases the number of last stage switches available to the first stage switch. According to the approach of the effective availability methods, the availability of last stage switches for calls of class c (appearing at the first stage switches) in multi-service switching networks can be calculated using the concept of the so-called equivalent network [11], determined individually for each of the classes of offered traffic. The assumption is that the equivalent single-service network for calls of class c services only the considered type of calls. The equivalent switching network has the same physical topology as the real multi-service network. The only difference is that the load (fictitious) of each inter-stage links have the load $y_{c,l}$ equal to the blocking probability calculated for calls of class c in the real, between stages l and $l+1$. This probability can be determined on the basis of appropriate models of inter-stage groups that will be presented further on in the article. To determine the effective availability $d_{c,3}$ in networks, the universal formula [11] for 3-stage switching networks with multi-service traffic can be used:

$$d_{c,3} = [1 - \pi_{c,3}]v + \pi_{c,3}y_{c,1} + \pi_{c,3}[v - y_{c,1}]y_{c,3}[1 - \pi_{c,3}], \tag{11}$$

where $\pi_{c,3}$ – probability of direct unavailability of a given (specified) switch in stage 3 for a call of class c; $\pi_{c,3}$ that determines the probability of an event in which a connection of class c cannot be set up between specified (given) switches of the first and the last stage [4]; $y_{c,l}$ – fictitious load of inter-stage link between stages l and $l+1$ in the equivalent switching network for calls of class c, equal in terms of the value to the blocking probability $[e_{c,l}]_f$ for calls of class c in the link of a real network [4].

Blocking Probability in Inter-stage Links. In the models that have been developed and used thus far, the influence of threshold mechanisms used in the outgoing directions on the blocking probability of the inter-stage links was determined by mapping the thresholds from outgoing directions (with higher capacity) to the thresholds in inter-stage links (with lower capacity) [7]. In the method described in this article, this influence is reflected by increasing the number of traffic classes that are offered to inter-stage links. The increase in the number of traffic classes results directly from the threshold mechanism applied to a given outgoing direction since the calls of class c, offered to outgoing directions, can be allocated ($q_c + 1$) different $t_{c,u}$ values of resources (AUs), depending on the occupancy state of the direction. Consequently, the simple full-availability model (FAG), without threshold mechanisms, can be used to model the inter-stage links with the increased number of traffic classes offered [6]. In order to avoid duplication of indexes, denoting particular traffic classes offered to the inter-stage links, the following formulas should be used:

$$\forall_{1 \leq c \leq m} \forall_{0 \leq u \leq q_c} t_{c+u+\sum_{z=1}^{c-1} q_z} = t_{c,u}, \; \forall_{1 \leq c \leq m} \forall_{0 \leq u \leq q_c} \mu_{c+u+\sum_{z=1}^{c-1} q_z} = \mu_{c,u}. \quad (12)$$

In order to calculate the occupancy distribution in full-availability groups (modelling inter-stage links with the increased number of traffic classes) it is necessary to calculate traffic intensity in particular threshold areas of the outgoing direction, which is offered subsequently to the interstage links. Let us assume that in the pre-threshold area of the outgoing direction the traffic intensity of the class c traffic stream that requires $t_{c,0}$ AUs is equal to $A_{c,0}$. Consequently, an inter-stage link will be offered a class c call that demands $t_{c+0} = t_{c,0}$ AUs to set up a connection, but the traffic intensity for this traffic stream will be equal to $A_{c,0}(1 - e_{c,1})/v$, where $e_{c,1}$ is the blocking probability for class c calls with the original demands (in the pre-threshold area). Subsequently, the traffic blocked in the pre-threshold area is offered to the first threshold area with the intensity $A_{c,0}e_{c,1}$. As a result, this traffic leads to the appearance of "new" streams that require $t_{c+1} = t_{c,1}$ in the inter-stage link. Continuing the above reasoning, the part of the traffic that is not blocked in the area u creates "new" traffic of class c with the value:

$$(1 - b_{c,u+1}) \frac{A_{c,0}}{v} \prod_{q=1}^{u} b_{c,q}, \quad (13)$$

where the parameter $b_{c,u}$ determines the blocking probability of calls of class c to which a given number of AUs is allocated, proper for the threshold area $u - 1$:

$$b_{c,u} = \sum_{n=Q_{c,u}^L+1}^{V_L} [Q_n]_{V_L}. \quad (14)$$

Note that in Formula (13) the traffic intensity, that demands $t_{c,u}$ AUs, is divided by v in order to take into account the difference in the capacity of an outgoing direction and an interstage links (an outgoing direction is v times higher than an inter-stage link).

With the values of Erlang, Enget and Pascal traffic offered to inter-stage links determined, on the basis of the above reasoning, we are in position to determine the occupancy distribution $[Q_n]_{V_F}$ in the inter-stage link with the capacity $V_F = f$ on the basis of Formula (9), assuming the value of the conditional transition coefficient to be equal to 1 and taking into account the increased number of traffic classes and capacity of the inter-stage links. This distribution allows the blocking probability for all the traffic classes offered to the inter-stage groups to be determined (Formula (10)).

The values of the blocking probability in the inter-stage links determined on the basis of the method presented in Sect. 3.2 make it possible to determine the effective availability parameter. Thus modified value of the effective availability provides then the basis for the modification of the two methods for a determination of the internal point-to-point blocking probability, i.e. the direct MDu method.

Direct Method. The internal point-to-point blocking probability in the universal direct method (MDu), based on the direct method (MD) [8], is defined as the ratio between the free links (for calls of class c) in the group of switches unavailable for a switch and all free (unoccupied) links that belong to a given direction. With the assumption that the occupancy probabilities of any links in the group are equal and independent of the occupancy of other links, the average value of the internal blocking probability can be expressed by the following formula:

$$E_{\mathrm{In}}(c) = (v - d_{c,3})/v, \tag{15}$$

where v is the capacity of the output direction, expressed in the number of links, whereas $d_{c,3}$ is the average value for availability (the number of available switches of the last stage) for a call of class c. This parameter can be determined on the basis of Formula (11), in which the parameters of the fictitious load $y_{c,l}$ are determined on the basis of the full-availability group model with an increased number of traffic classes.

4   Numerical Results

The generalized MDu method for determining the blocking probability in switching networks with resource allocation control mechanisms is an approximate method. To evaluate its accuracy and the adopted assumptions for the method, the results of the analytical calculations were compared with the data obtained in simulation experiments. The simulation study was carried out for 3-stage Clos networks. The switching network under investigation was composed of square switches $v \times v$ links, each with the capacity of f AUs. The data obtained on the basis of the simulation study are presented in Fig. 2 as points with the confidence intervals calculated after the t-Student distribution (with 95-percent confidence level) for 5 series with 1,000,000 calls each (the classes with the lowest call intensity). In a large number of instances, the value of the confidence interval is lower that the height of the symbol representing the simulation result.

A comparison was also made for the changes in the blocking probability for individual call classes obtained on the basis of the proposed generalized MDu method and on the basis of its original version. This is illustrated in Fig. 3 that show the changes in the values of errors introduced by the modified MDu method and the values of errors from before the introduction of modifications. Due to a limited length of the present article, the results of the study are limited only to one network with the following structure: $v = 4$, $f = 42$ AUs, $V = 168$ AUs. This network was offered traffic with the following parameters: Traffic classes: $m = 4$, $t_{1,0} = 1$ AU, $\mu_{1,0}^{-1} = 1$, $t_{2,0} = 6$ AUs, $\mu_{2,0}^{-1} = 1$, $t_{3,0} = 8$ AUs, $\mu_{3,0}^{-1} = 1$, $t_{4,0} = 12$ AUs, $\mu_{4,0}^{-1} = 1$, $t_{4,1} = 8$ AUs, $\mu_{4,1}^{-1} = 1.25$, $t_{5,0} = 12$ AUs, $\mu_{5,0}^{-1} = 1$, $t_{5,1} = 8$ AUs, $\mu_{5,1}^{-1} = 1.5$; Threshold mechanism: $q_4 = 1$, $q_5 = 1$, $Q_{4,1} = Q_{5,1} = 126$ AUs.

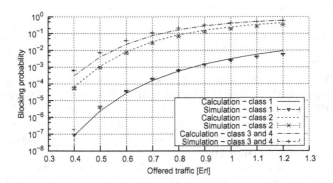

Fig. 2. Total blocking probability according to direct MBu method

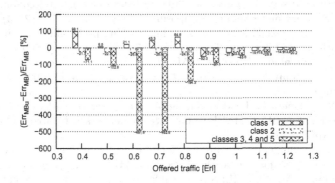

Fig. 3. Change in the blocking probability error (Err) – methods MD and MDu

5   Conclusions

This article proposes a modification to available methods for determining the point-to-point blocking probability that is based on the concept of a substitution of the system of inter-stage links with threshold mechanisms for a system of full-availability groups with appropriately increased number of traffic classes, which in consequence leads to considerably higher accuracy of the latter methods. In the case of the MDu method, the increase in the accuracy is particularly distinct for those traffic classes that demand larger amounts of resources[1].

References

1. Bonald, T.: A recursive formula for estimating the packet loss rate in IP networks. In: Proceedings of the Fourth International ICST Conference on Performance Evaluation Methodologies and Tools, pp. 56:1–56:2 (2009)
2. Chrysos, N., Minkenberg, C., Rudquist, M., Basso, C., Vanderpool, B.: SCOC: high-radix switches made of bufferless clos networks. In: 2015 IEEE 21st International Symposium on High Performance Computer Architecture (HPCA), pp. 402–414 (2015)
3. Ershov, V.: Some further studies on effective accessibility: fundamentals of tele-traffic theory. In: Proceedings of 3rd International Seminar on Teletraffic Theory, pp. 193–196. Moscow (1984)
4. Głąbowski, M., Sobieraj, M.: Analytical modelling of multiservice switching networks with multiservice sources and resource management mechanisms. Telecommun. Syst. **66**(3), 559–578 (2017)
5. Głąbowski, M.: Recurrent calculation of blocking probability in multiservice switching networks. In: Proceedings of Asia-Pacific Conference on Communications, pp. 1–5, Busan (2006)
6. Głąbowski, M., Kaliszan, A., Stasiak, M.: Modeling product-form state-dependent systems with BPP traffic. Perform. Eval. **67**, 174–197 (2010)
7. Głąbowski, M., Sobieraj, M.: Modelling of network nodes with threshold mechanisms and multi-service sources. In: 2014 16th International Telecommunications Network Strategy and Planning Symposium (Networks), pp. 1–7 (2014)
8. Głąbowski, M., Stasiak, M.: Point-to-point blocking probability in switching networks with reservation. Ann. Telecommun. **57**(7–8), 798–831 (2002)
9. Pras, A., Nieuwenhuis, L., van de, R.M., Mandjes, M.: Dimensioning network links: a new look at equivalent bandwidth. IEEE Netw. **23**(2), 5–10 (2009). http://doc.utwente.nl/65443/
10. Sehery, W., Clancy, T.: Load balancing in data center networks with folded-Clos architectures. In: 2015 1st IEEE Conference on Network Softwarization (NetSoft), pp. 1–6, April 2015
11. Stasiak, M.: Combinatorial considerations for switching systems carrying multi-channel traffic streams. Anna. Des Télécomm. **51**(11–12), 611–625 (1996)
12. Xia, Y., Hamdi, M., Chao, H.: A practical large-capacity three-stage buffered Clos-network switch architecture. IEEE Trans. Parallel Distrib. Syst. **27**(2), 317–328 (2016)
13. Ye, T., Lee, T., Hu, W.: AWG-based non-blocking Clos networks. IEEE/ACM Trans. Netw. **23**(2), 491–504 (2015)

---

[1] The work is supported by Polish Ministry of Science and Higher Education 08/82/DSPB/8216.

Inconsistencies Among Spectral Robustness Metrics

Xiangrong Wang[1(✉)], Ling Feng[2], Robert E. Kooij[1,3], and Jose L. Marzo[4]

[1] Faculty of Electrical Engineering, Mathematics and Computer Science,
Delft University of Technology, Delft, The Netherlands
`x.wang-2@tudelft.nl`
[2] Computing Science Department, Institute of High Performance Computing,
A*STAR, Singapore, Singapore
`fengl@ihpc.a-star.edu.sg`
[3] iTrust, Centre for Research in Cyber Security,
Singapore University of Technology and Design, Singapore, Singapore
`robert_kooij@sutd.edu.sg`
[4] Institute of Informatics and Applications, University of Girona,
Girona, Spain
`joseluis.marzo@udg.edu`

Abstract. Network robustness plays a critical role in the proper functioning of modern society. It is common practice to use spectral metrics, to quantify the robustness of networks. In this paper we compare eight different spectral metrics that quantify network robustness. Four of the metrics are derived from the adjacency matrix, the others follow from the Laplacian spectrum. We found that the metrics can give inconsistent indications, when comparing the robustness of different synthetic networks. Then, we calculate and compare the spectral metrics for a number of real-world networks, where inconsistencies still occur, but to a lesser extent. Finally, we indicate how the concept of the R^*-value, a weighted sum of robustness metrics, can be used to resolve the found inconsistencies.

Keywords: Inconsistency · Graph theory · Network theory · Graph spectra · Robustness metrics

1 Introduction

Failures of real-world networks, such as blackouts in power grids, traffic congestion in transportation networks, and economic crisis in economic networks, can have an enormous impact on society, in terms of costs, safety and disruption [14]. Therefore, understanding the robustness of networks, which reflects the extent to which the networks can maintain their functionality under perturbations imposed upon them, is crucial for modern critical infrastructures. Quantifying the robustness of networks enables us to design, optimize and control the networks.

T. Q. Duong et al. (Eds.): Qshine 2018, LNICST 272, pp. 119–136, 2019.
https://doi.org/10.1007/978-3-030-14413-5_10

Recent advances in the field of network science present a number of robustness metrics, both from the topological domain and the spectral domain, characterizing the structural and dynamical properties of networks. Examples are degree distribution reflecting the connectivity of a network [1], modularity for the community structure [17], spectral radius [28] characterizing the virus spread in a network, and the algebraic connectivity [7,27] which relates to the synchronization of networks of coupled oscillators [25]. However, there is a lack of study on the relation between the robustness metrics and the interpretation of each metric in terms of the network robustness. In [19], it is shown that most metrics are not mutually independent, indicating redundancy in the characterization of robustness.

Spectral graph theory is applied in various aspects of complex networks, see for example surveys by Cvetković [5,6]. Particularly, eigenvalues and eigenvectors are used for the analysis of the robustness of complex networks [12,28,31,33]. In this paper we focus on the quantification of the robustness of complex networks, by means of spectral metrics [27]. Our main contribution is showing the occurrence of inconsistencies among the spectral metrics that quantify robustness. The inconsistencies mean that for a pair of graphs, say G and H, a pair of robustness metrics $\{M_1, M_2\}$ point in opposite direction, i.e. according to metric M_1 the graph G is more robust, but according to the metric M_2 the graph H is more robust. Because we consider eight different spectral metrics, we need to construct inconsistencies among 28 pairs of metrics. We will realize this number of inconsistencies with the help of 10 graphs, all having $N = 7$ nodes and $L = 10$ links. Next we show that inconsistencies also occur for arbitrary large pairs of graphs. Then, we calculate and compare the spectral metrics for a number of real-world networks, with numbers of nodes and links in the range 21–29 and 22–37, respectively. Finally, we indicate how the concept of the R^*-value, a weighted sum of robustness metrics, can be used to resolve the found inconsistencies.

2   Spectral Robustness Metrics

In this section, we present the definitions of the eight robustness metrics and their relation to the robustness of networks. Let $G(N, L)$ be an undirected graph with N nodes and L links. The adjacency matrix A of a graph G is an $N \times N$ symmetric matrix with elements a_{ij} that are either 1 or 0 depending on whether or not there is a link between nodes i and j. The eigenvalues of $A = A^T$ are real and can be ordered as $\lambda_N \leq \lambda_{N-1} \leq \ldots \leq \lambda_1$.

Another graph related matrix is the Laplacian matrix $Q = \Delta - A$, where $\Delta = \text{diag}(d_i)$ is the $N \times N$ diagonal degree matrix and the degree of node i is $d_i = \sum_{j=1}^{N} a_{ij}$. The eigenvalues of Q are non-negative and at least one of them is zero [27]. The eigenvalues of Q can be ordered as $0 = \mu_N \leq \mu_{N-1} \leq \ldots \leq \mu_1$.

We first present four robustness metrics which are based upon the adjacency spectrum.

2.1 Spectral Radius (SR)

The spectral radius [27] refers to *the largest eigenvalue* λ_1 of the adjacency matrix of a graph

$$SR = \lambda_1. \tag{1}$$

A larger spectral radius is associated with higher robustness of the networks with respect to link/node removals.

2.2 Spectral Gap (SG)

The spectral gap is expressed as

$$SG = \lambda_1 - \lambda_2. \tag{2}$$

According to the Perron-Frobenius theorem, λ_1 of a graph is always positive. The largest spectral gap $\lambda_1 - \lambda_2 = N$ occurs in the case of a complete graph. The spectral gap plays an important role in the dynamic processes on graphs [27]. The larger the spectral gap is, the higher the robustness of a network. A network with a large spectral gap is typically onion structured which is more robust against malicious attacks and random removals [34].

2.3 Natural Connectivity (NC)

Natural connectivity [10,16] is defined as

$$NC = \ln\left(\frac{1}{N}\sum_{k=1}^{N} e^{\lambda_k}\right). \tag{3}$$

where λ_k is the kth eigenvalue of the adjacency matrix of a graph. The natural connectivity is proposed as a spectral measure for the robustness of complex networks in terms of the redundancy of alternative paths [16,33]. The higher the natural connectivity is, the higher the robustness of a network.

2.4 Minimum-Maximum Eigenvalue Ratio (MM)

The ratio of the maximum eigenvalue λ_1 to the minimum eigenvalue λ_N is defined as [36]

$$MM = \left|\frac{\lambda_1}{\lambda_N}\right|. \tag{4}$$

The ratio is used in signal detection and the stability of neural networks [22, 36]. The higher the ratio of the maximum to the minimum is, the higher the robustness of a network.

Next, we introduce four spectral metrics based upon the Laplacian spectrum.

2.5   Algebraic Connectivity (AC)

The algebraic connectivity, coined by Fiedler [13], refers to the second smallest eigenvalue of the Laplacian matrix Q

$$AC = \mu_{N-1}. \tag{5}$$

It has been shown [15] that the larger the algebraic connectivity is, the more difficult it is to cut the network into components, and the higher the robustness of a network is [2, 32].

2.6   Number of Spanning Trees (NST)

The total number of spanning trees [27] can be written in terms of the eigenvalues of the Laplacian matrix as

$$NST = \frac{1}{N} \prod_{j=1}^{N-1} \mu_j. \tag{6}$$

The total number of spanning trees is suggested as an indicator of network robustness [4, 9]. The higher NST is, the higher the robustness of a network is.

2.7   Effective Graph Resistance (EGR)

The effective graph resistance is determined by

$$EGR = N \sum_{k=1}^{N-1} \frac{1}{\mu_k}, \tag{7}$$

where μ_k is the kth eigenvalue of the Laplacian matrix of a graph. The smaller the effective graph resistance is, the higher the robustness of a network is [8, 30, 31].

2.8   Eigenvalue Ratio (ER)

The eigenvalue ratio refers to the ratio of the second smallest eigenvalue μ_{N-1} to the largest eigenvalue μ_1 of the Laplacian matrix Q of a graph

$$ER = \frac{\mu_{N-1}}{\mu_1}. \tag{8}$$

The eigenvalue ratio is used to characterize the synchronizability of networks. If the eigenvalue ratio is larger, a network exhibits a better synchronizability [3, 7, 20].

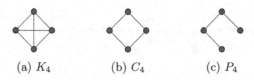

(a) K_4 (b) C_4 (c) P_4

Fig. 1. Three types of graphs with 4 nodes.

Table 1. Spectral robustness metrics for three graphs with 4 nodes: complete graph K_4, cycle graph C_4 and path graph P_4. All of the metrics indicate that K_4 is the most robust graph.

Graphs	SR	SG	NC	MM	AC	NST	EGR	ER
K_4	3	4	1.667	3	4	16	3	1
C_4	2	2	0.868	1	2	4	5	0.5
P_4	1.618	1	0.647	1	0.586	1	10	0.172

3   Evaluation of the Spectral Metrics

First we illustrate the use of the robustness metrics, by applying them to three simple networks on 4 nodes: K_4: a complete graph, C_4: a cycle graph, P_4: a path graph, see Fig. 1.   From Table 1, we see that the 8 different metrics, all rank the robustness of the three graphs, in the same way. For instance, all metrics indicate that K_4 is the most robust graph among the three graphs, as expected. This shows that, for the simple graphs here with only four nodes, the robustness metrics are consistent in their indications.

3.1   Ten Example Graphs

We use 10 different networks $G1 - G10$, to analyze the consistency of the 8 different metrics. For each of the networks, there are 7 nodes and 10 links, so that the differences are only in the way the links are constructed. Figure 2 presents the visual representations of the 10 networks. For the 10 networks, we determine the values of the 8 metrics described in the previous section. The results are listed in Table 2. It is clear that different metrics may give different indications as to which graph is the most robust, although each network has the same number of nodes and links. Such inconsistencies do not occur for the previous example of the simple graphs on four nodes.

To further illustrate this, we cross-compare each pair of metrics among the 8 metrics, and identify the pairs that give inconsistencies in Table 3. The pair of graphs in each cell are the graphs that lead to inconsistent indications of their relative robustness given by the two different metrics. For example, the cell that cross-compares the metrics Natural Connectivity (NC) and Algebraic Connectivity (AC), contains the graph pair $G1$ and $G3$. Indeed, according to Table 2, the NC indicates that $G1$ if more robust than $G3(1.51 > 1.44)$, while

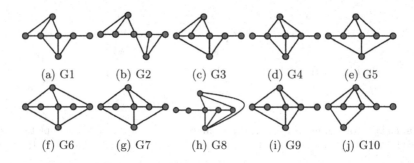

Fig. 2. The visualization of 10 artificial networks with 7 nodes and 10 links each.

AC indicates that $G3$ is more robust than $G1(0.67 > 0.63)$. For each pair of metrics, there is always at least one pair of graphs that are inconsistent, as seen in the Table 3. This means the inconsistencies are prevalent in the graphs studied, despite their similarities in terms of number of nodes and links.

3.2 All Connected Graphs with 7 Nodes and 10 Links

There are 132 possible non-isomorphic connected graphs with 7 nodes and 10 links. These graphs are generated from the programs called nauty and Traces [24]. We evaluate the inconsistency of the 8 spectral robust metrics for all the possible 132 graphs.

Figure 3 shows the rank of robustness for all the 132 graphs according to the 8 spectral robust metrics. The increase of ranking number means the decrease of robustness. The most robust graph has a ranking number 1.

The high variability in Fig. 3 suggests the inconsistency of the 8 spectral robust metrics when identifying the rank order for the 132 graphs. For example,

Table 2. Spectral robustness metrics, R* (P = 3). Every graph has $N = 7$ nodes and $L = 10$ links.

Gr	SR	SG	NC	MM	AC	NST	EGR	ER	R*
G1	3.21	1.74	1.51	**1.67**	0.63	55	21.67	0.12	0.697
G2	3	1	1.45	1.5	0.59	64	21.50	0.11	0.683
G3	3.12	1.68	1.44	1.43	0.67	64	20.56	0.12	0.681
G4	**3.35**	**2.35**	**1.58**	1.49	0.70	45	22.40	0.12	0.667
G5	3.01	1.92	1.37	1.23	**1.38**	95	16.32	**0.25**	0.649
G6	2.96	1.96	1.30	1	**1.38**	105	**15.62**	0.23	**0.715**
G7	2.98	1.98	1.33	1.16	1.33	101	15.88	0.23	0.698
G8	3.30	2.07	1.56	1.51	0.44	45	25.73	0.08	0.708
G9	3.16	2.16	1.44	1.25	0.83	69	18.99	0.14	0.662
G10	3.20	2.07	1.48	1.44	0.69	61	20.49	0.12	0.691

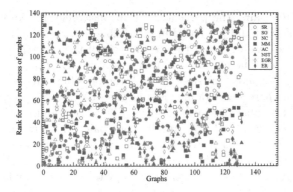

Fig. 3. Rank for the robustness of 132 graphs with 7 nodes and 10 links.

the most robust graph (rank 1, red line in Fig. 3) varies according to different spectral metrics. The spectra radius λ_1 and the natural connectivity NC identify graph 5 as the most robust network. The spectra gap SG and the algebraic connectivity AC rank graph 2 as the top robust network. According to the Minimum-Maximum eigenvalue detection MM, graph 56 is ranked the top. The number of spanning trees NST suggests that graph 124 is the most robust network. The effective graph resistance EGR identifies graph 95 and the eigenvalue ratio ER shows graph 129 as the most robust one.

Table 3. The comparison of spectral robustness metrics. For each pair of metrics, there is at least one pair of graphs that leads to inconsistent indication with respect to their relative robustness.

	SR	SG	NC	MM	AC	NST	EGR	ER
SR	X	G1:G5	G3:G9	G1:G4	G1:G3	G1:G2	G1:G2	G1:G3
SG	X	X	G1:G5	G1:G4	G3:G8	G1:G2	G1:G2	G3:G8
NC	X	X	X	G1:G4	G1:G3	G1:G2	G1:G2	G1:G3
MM	X	X	X	X	G1:G3	G1:G2	G1:G2	G1:G3
AC	X	X	X	X	X	G1:G2	G1:G2	G6:G7
NST	X	X	X	X	X	X	G3:G10	G1:G2
EGR	X	X	X	X	X	X	X	G1:G2
ER	X	X	X	X	X	X	X	X

For the pairs of graphs, the inconsistency of robust metrics is presented in Table 4 and Fig. 4. Table 4 shows the percentage of inconsistency among all possible pairs of graphs. For all the 132 graphs, there are $\binom{132}{2} = 8646$ possible graph pairs. For a pair of graphs, one robust metric concludes which graph is more robust than the other one. Two robust metrics provide either consistent conclusion or inconsistent conclusion for the same pair of graphs. After going through all the 8646 graph pairs, the percentage of inconsistency is computed for each pair of robust metrics and presented in each table cell. In Table 4, the minimum percentage is 0.06 resulted from the metric pair of the spectral radius and the natural connectivity. The non-zero minimum inconsistency percentage indicates that there is no complete consistency for all the possible pairs of graphs.

The maximum percentage of inconsistency is 91% between metrics of natural connectivity and number of spanning trees. The second and third top inconsistency percentages, 89% and 88% result from pairs of robust metrics (EGR, NST) and (EGR, ER). The top three percentages of inconsistency are further presented in Fig. 4. The high percentages (higher than 88%) highlight the challenges for graph designer to design a completely robust topology.

Table 4. The percentage of inconsistency between pairs of metrics.

	SR	SG	NC	MM	AC	NST	EGR	ER
SR	X	26%	6%	27%	68%	88%	19%	72%
SG	X	X	32%	46%	49%	64%	4%	53%
NC	X	X	X	23%	72%	91%	15%	75%
MM	X	X	X	X	67%	73%	26%	67%
AC	X	X	X	X	X	20%	86%	8%
NST	X	X	X	X	X	X	89%	16%
EGR	X	X	X	X	X	X	X	88%
ER	X	X	X	X	X	X	X	X

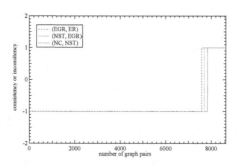

Fig. 4. The consistency (marked as 1) and inconsistency (marked as −1) results for metrics pairs of (NC, NST), (EGR, NST) and (EGR, ER).

3.3   Arbitrary Large Graphs

The examples given in the previous sections are for small graphs, with only 7 nodes. We will now give an example of an inconsistency occurring for a family of pairs of graphs that can be arbitrary large. We first consider a complete bi-partite graph K_{N_1,N_2}, consisting of two disjoint sets S_1 and S_2, containing, respectively, N_1 and N_2 nodes, such that all nodes in S_1 are connected to all nodes in S_2, while within each set no connections occur. According to [5], the spectral radius and algebraic connectivity for K_{N_1,N_2} satisfy $SR(K_{N_1,N_2}) = \sqrt{N_1 N_2}$ and $AC(K_{N_1,N_2}) = min\{N_1, N_2\}$, respectively. Note that K_{N_1,N_2} has $N_1 + N_2$ nodes.

The second graph we consider is the windmill graph $W(\eta, k)$, which consists of η copies of the complete graph K_k, with every node connected to a common node, see Fig. 5. Note that $W(\eta, k)$ has $\eta k + 1$ nodes.

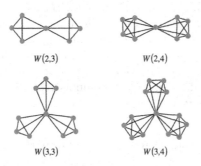

$w(2,3)$ $w(2,4)$

$w(3,3)$ $w(3,4)$

Fig. 5. Illustration of some windmill graphs $W(\eta, k)$

Estrada [11] has shown that the algebraic connectivity of $W(\eta, k)$ satisfies $AC(W(\eta, k)) = 1$, while the spectral radius $SR(W(\eta, k))$ is given by the largest zero of

$$f(\lambda) = \lambda^2 - (k - 1)\lambda - \eta k = 0. \tag{9}$$

We now construct an inconsistency by choosing $N_1 = 2$ and $\eta = 2$. Assuming that the number of nodes for both graphs are equal, it follows that $N_2 = 2k - 1$. So we consider the family of pairs of graphs $H_1 = K_{2,2k-1}$ and $H_2 = W(2, k)$, with $k > 1$. It follows from the properties mentioned above that $AC(H_1) = 2 > 1 = AC(H_2)$. Substitution of $\lambda = k$ and $\eta = 2$ into Eq. (9) gives $f(k) = -k < 0$. Therefore the spectral radius of H_2 is larger than k. On the other hand, $SR(H_1) = \sqrt{2(2k - 1)}$ which is smaller than k for $k \geq 4$. Hence, $SR(H_1) < SR(H_2)$, implying an inconsistency for the graphs $\{H_1, H_2\}$ for the pair of metrics $\{SR, AC\}$, for every $k \geq 4$.

For the inconsistency constructed above, the two graphs H_1 and H_2 have the same number of nodes, but not the same number of links. However, it is possible to construct inconsistencies for pairs of graphs with the same number of nodes and links. One could use a windmill graph $W(\eta, k)$ and an Erdös-Rényi graph $ER(N, L)$, choosing N and L such that the two graphs have the same number of nodes and links. For instance, consider $H_3 = W(10, 10)$ and $H_4 = ER(101, 550)$. Then both graphs have 101 nodes and 550 links. For one realization of $ER(101, 550)$ we obtain $AC(H_3) = 1 < 3.88 = AC(H_4)$ while $SR(H_3) = 15.47 > 11.66 = SR(H_4)$, which implies an inconsistency. It is possible to generalize this example to families of pairs of graphs of arbitrary size also.

3.4   Real-World Networks

In addition to artificial graphs, we now compare the different metrics on 6 real-world networks to analysis their consistencies. These networks are taken from the so-called Internet Topology Zoo (http://www.topology-zoo.org/), a collection of data network topologies from around the world, see also [18]. The networks include AboveNet, AGIS, Atmnet, Bbnplanet, Bizent and BtEurope networks. Their topologies are presented in Fig. 6. These networks have comparable number of nodes N and number of links L, but differ significantly in their topological structures, as shown. Table 5 shows the robustness metrics of the various networks, and Table 6 shows the relative rankings of them according to each metric. In both tables, the results depicted in columns R*5 (for P = 5) and R*10 (for P = 10) are analyzed in Sect. 5.

Table 5. The spectral robustness, R*5 (P=5) and R*10 (P=10) metrics for real-world networks.

Gr	N	L	SR	SG	NC	MM	AC	NST	EGR	ER	R*5	R*10
Abovenet	23	31	3.13	0.35	1.21	1.20	0.17	262K	425	0.027	0.776	0.638
AGIS	25	30	3.17	0.58	1.11	1.13	0.20	6376	579	0.027	0.768	0.668
Atmnet	21	22	2.29	0.28	0.87	1.00	0.09	107	592	0.019	0.742	0.572
Bbnplanet	27	28	2.89	0.42	0.95	1.05	0.12	34	931	0.018	0.762	0.654
Biznet	29	33	2.50	0.09	0.94	1.09	0.04	8856	1391	0.008	0.749	0.667
BtEurope	24	37	5.10	2.56	2.15	1.49	0.44	612K	372	0.031	0.799	0.665

Table 6. The relative rankings of real-world networks according to each spectral robustness metrics.

Graphs	SR	SG	NC	MM	AC	NST	EGR	ER	R*5	R*10
AboveNet	3	4	2	2	3	2	2	2	2	5
AGIS	2	2	3	3	2	4	3	3	3	1
Atmnet	6	5	6	6	5	5	4	4	6	6
Bbnplanet	4	3	4	5	4	6	5	5	4	4
Biznet	5	6	5	4	6	3	6	6	5	2
BtEurope	1	1	1	1	1	1	1	1	1	3

The same inconsistencies are found in these 6 networks, although the relative rankings do not differ much. In particular, the BtEurope network has the highest robustness ranking for every metric. In addition, we see that EGR and ER metrics give exactly the same rankings for all of the networks. Therefore, it is interesting to note that, although the various metrics could give different indications of relative robustness rankings, the most robust structure in real-world networks can be consistent across different metrics. It shows a certain level of consistency, with inconsistencies remaining in the less robust network structures.

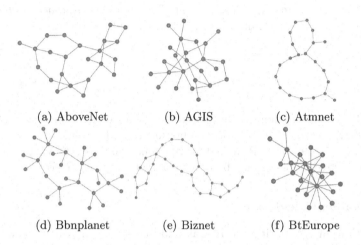

(a) AboveNet   (b) AGIS   (c) Atmnet

(d) Bbnplanet   (e) Biznet   (f) BtEurope

Fig. 6. The visualization of six real-world networks.

4 R^*-Value as Robustness Metric

One possible approach to deal with the observed inconsistencies, is to make explicit, for every specific case study, the definition of robustness. For instance, Wang et al. [29] studied the robustness of 33 metro networks in the world, and took as an experimental robustness definition, the fraction of nodes that have to be removed from the metro network, such that the remaining largest connected component is 90% of the original network size. It is shown in [29] that with this definition, the effective graph resistance captures very well the robustness of the metro networks.

Another way to deal with the inconsistencies is to use the information of all spectral metrics. As an example, based upon Table 2, we might conclude that graphs $G4$ and $G6$ are the most robust, because for both, 3 out of the 8 spectral metrics indicate them as the most robust. Of course one could construct much more complicated ways to combine the 8 spectral metrics.

4.1 R^*-Value: Definition

The idea of expressing the robustness value (or R-value) of a network as a weighted sum of a number of metrics, was first proposed by Trajanovski et al. [26]:

$$R = \sum_{k=1}^{K} s_k \, t_k, \tag{10}$$

where K is the number of considered metrics, t_k denote graph metrics and s_k their corresponding weights. However, this is a static analysis and the problem of determining the weights s_k is still present. To overcome this issue, the R-value concept was enhanced by Manzano et al. in [21], leading to the concept of the R^*-value, see Eq. (11):

$$R^* = \sum_{k=1}^{K} \widehat{v}_k \, t_k. \tag{11}$$

Here, the weights $\widehat{v}_k$ reflect the relative importance of the metrics t_k, when elements of the network are removed subsequently. The values of the weights are determined by applying Principal Component Analysis (PCA). In the next subsection we give more details about the calculation of the R^*-value.

4.2 R^*-Value: Calculation

The algorithm for calculating the R^*-value for a network, depends on two integers, denoted by P and M. Here P denotes the maximum number of network elements that are removed from the original network. For every number p of removed network elements, with $1 \leq p \leq P$, we conduct M independent experiments, in which each of the K metrics are determined.

The sequence of calculations is shown in Fig. 7 and detailed below. First, the metrics of the initial network with no attacks (i.e. $p = 0$) are calculated providing K metrics measurements. Note that this result is the same for all M experiments since the topology always remains the same (first row in the metrics matrix, see Fig. 7).

Once the list of elements to be removed is obtained, the K metrics are calculated for all $P \times M$ pairs. This provides a $(P \times M \times K)$ metrics matrix which contains all the computed metrics. Then, the correlation matrix $(K \times K)$ of all metric results is obtained. Then PCA is applied to the correlation matrix obtaining the v weights of each metric. PCA provides a K-dimensional eigenvector, the larger eigenvalue and its corresponding eigenvalue is selected. For comparison purposes, the initial value of R^* is normalized to 1 (maximum robustness) and the weights are modified accordingly.

Finally, by multiplying the $\widehat{v}_k$ weights for all rows in the $(P \times M \times K)$ metrics matrix as indicated in Eq. (11), the normalized robustness value R^* can be computed for all $P \times M$ cases. Then the robustness of a network is the average of all R^* values.

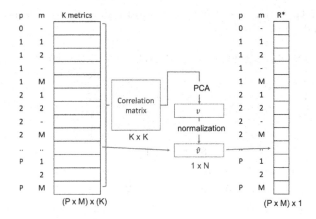

Fig. 7. $R*^-$ value calculation overview

In the section we will use the concept of the R^*-value as the "ground truth" for the robustness of networks, in order to resolve the inconsistencies mentioned in Sect. 3.

Following the insights obtained in [23], we base the R^*-value on a combination of 10 relevant metrics. Average Nodal Degree (AND), Efficiency (EFF) and Spectral Radius (SR) are the representative structural metrics, while Largest Connected Component (LCC) and Average Two Terminal Reliability (ATTR) are the representative structural metrics for fragmentation. Algebraic Connectivity (AC) and Natural Connectivity (NC) for represent connectivity and, finally, Closeness Centrality (CLC) and Eigenvector Centrality (EC) represent centrality properties. The R^*-value is evaluated under a random node removal strategy.

5 R^*-Value Versus Spectral Metrics: Results

5.1 Comparison for the 10 Example Graphs $G1 - G10$

For this set of results, R^* has been computed by randomly removing up to 3 nodes (P = 3) in 20 independent samples (M = 20). The results are shown in column R^* of Table 2.

If we use the obtained R^*-values as ground truth, then we can resolve the inconsistencies reported in Table 3. For instance, the graph pair $G1 : G5$ leads to an inconsistency for the metric pair SR and SG. Because $R^*(G1) > R^*(G5)$ we resolve the inconsistency by stating that $G1$ is more robust than $G5$. In this way we can resolve Table 3 completely. Note that it is difficult to draw generic conclusions from this, because Table 3 only contains 8 graph pairs, namely, $\{(G1 : G5), (G3 : G9), (G1 : G4), (G1 : G3), (G3 : G8), (G1 : G2), (G3 : G10), (G6 : G7)\}$. To obtain generic conclusions one could study all resolved inconsistencies for the 8646 graph pairs mentioned in Sect. 3.2. This approach is left for further study.

Table 7. The relative rankings of artificial networks according to each spectral robustness metrics and R^* (P = 3).

Graphs	SR	SG	NC	MM	AC	NST	EGR	ER	R*
G1	3	8	3	1	8	8	8	5	4
G2	8	10	5	3	9	5	7	9	6
G3	6	9	6	4	7	5	6	5	7
G4	1	1	1	3	5	9	9	5	8
G5	7	7	8	6	1	3	3	1	10
G6	10	6	10	8	1	1	1	2	1
G7	9	5	9	7	3	2	2	2	3
G8	2	3	2	9	10	9	10	10	2
G9	5	2	6	5	4	4	4	4	9
G10	4	3	4	2	6	7	5	5	5

Table 7 presents the robustness ranking of the ten graphs G1–G10, according to the spectral metrics and the R^*-value. Here, rank 1 denotes the most robust network, while rank 10 denotes the least robust network. If we take the R^*-value as the ground truth for the robustness ranking, a few observations can be made from Table 7. Firstly, if we are only interested in the most robust network, then AC, NST and EGR lead to the same network as R^*, namely G6. However, the second most robust network according to R^*, i.e. G8, is ranked very low by these three spectral metrics, namely 10th, 9th and 10th, respectively. Secondly, the least robust network according to R^*, i.e. G5, is never ranked as the least robust network by any of the 8 spectral metrics. The closest is metric NC, which gives G5 rank 8. Finally, out of the 10 considered graphs, the rankings of the spectral metrics re the most consistent with that of R^*, for G10. In contrast G5 is the least consistent with R^*.

5.2 Comparison for the 6 Real-World Networks

In this section we compare the robustness ranking of the six real-world networks introduced in Sect. 3.3, according to the spectral metrics and the R^*-value. We will consider two scenarios for the computation of the R^*-value, namely removal up til 5 nodes (P = 5) and up til 10 nodes (P = 10). The number of independent samples remains $M = 20$. We denote the resulting R^*-values by R^*5 and R^*10, respectively. The two right-most columns Table V give the values of R^*5 and R^*10. Table 7 gives the corresponding rankings for the real-world networks.

A few observations can be made from these tables. First of all, the two scenarios $P = 5$ and $P = 10$ lead to different rankings. This is not surprising because $P = 10$ corresponds to a more severe attack than $P = 5$. Secondly, R^*5 states that BtEurope is the most robust network. This is in line with the ranking of all spectral metrics. The least robust network according to R^*5, i.e.

Atmnet, is also recognized as a vulnerable network by the other metrics, with three metrics denoting it is the most vulnerable network.

Table 8. The distances of the rankings of real-world networks to R^*5 rankings.

Graphs	SR	SG	NC	MM	AC	NST	EGR	ER
AboveNet	1	2	0	0	1	0	0	0
AGIS	1	1	0	0	1	1	0	0
Atmnet	0	1	0	0	1	1	2	2
Bbnplanet	0	1	0	1	0	2	1	1
Biznet	0	1	0	1	1	2	1	1
BtEurope	0	0	0	0	0	0	0	0
Total distance	2	6	0	2	4	6	4	4

Table 9. The distances of the rankings of real-world networks to R^*10 rankings.

Graphs	SR	SG	NC	MM	AC	NST	EGR	ER
AboveNet	2	1	3	3	2	3	3	3
AGIS	1	1	2	2	1	3	2	2
Atmnet	0	1	0	0	1	1	2	2
Bbnplanet	0	1	0	1	0	2	1	1
Biznet	3	4	3	2	4	1	4	4
BtEurope	2	2	2	2	2	2	2	2
Total distance	8	10	10	10	10	12	14	14

Tables 8 and 9 present the distances between spectral metrics and R^* obtained by comparing the rankings of real-world networks presented in Table 6. The smaller the distance, the more similar the ranking is to the ranking of R^*5 and R^*10, respectively. The first conclusion is that, as expected, the rankings of the spectral metrics are the more similar, when the number of removed elements P is smaller. This makes sense as spectral metrics analyze the initial network (i.e. $P = 0$). For instance, Table 8 shows that NC gives exactly the same ranking as R^*5, i.e. the sum of distances equals 0. Similarly, SR and MM also present quite similar rankings (distances = 2). Instead, when comparing this with R^*10, accumulated distances are always larger (ranging from 8 (SR), best case, to 14 (EGR and ER), worst case).

Considering both tables, NC and RC complement each other, being quite similar to the R^* ranking both for small and large removal of nodes. On the other hand, NST, EGR and ER lead to less similarity in rankings, in most of the cases.

From the perspective of the network topologies, it is interesting to note that for BtEurope all spectral metrics provide very accurate results for small removal of nodes ($P = 5$), while Atmnet is the most consistent for $P = 10$.

6   Conclusion

We have shown that among 8 of frequently used spectral metrics, inconsistencies occur when using them to capture the robustness of networks. The non-zero and high percentages of inconsistency, for pair of graphs from the set of 132 graphs with 7 nodes and 10 links, suggest the challenge for the complete robust quantification of graphs. Such inconsistency is more pronounced in the artificial networks we generated than in the real-world networks tested.

One possible approach to deal with the inconsistencies, is to make explicit, for every specific case study, the definition of robustness, as was done by Wang et al. [29], who studied the robustness of 33 metro networks in the world. Another way to deal with the inconsistencies would be to use the information of all spectral metrics. With enough data at hand and with a baseline for explicit experimental values for the robustness, such as in [29], this line of reasoning, seems worth pursuing. This merging of network science and machine learning has recently also been suggested by Zanin et al. [35].

In this paper we resolved the inconsistencies by considering the so-called R^*-value, see [21], as the ground truth for robustness. In Sect. 5, Table 8 shows that robustness ranking according to spectral metrics is more similar to ranking according to R^*, when the number of removed elements (P) is small. This makes sense as spectral metrics analyze the initial network (i.e. $P = 0$). In particular, Natural Connectivity (NC) gives precisely the same ranking as R^* for $P = 5$, while Spectral Radius (SR) and Minimum-Maximum eigenvalue ratio (MM) are also good approximations. Comparisons for larger amounts of node removals ($P = 10$), show that spectral metrics generally give less similar rankings, see Table 9. When looking at the network topologies, all spectral metrics provide similar results for BtEuropa upon removal of small number of nodes and for Atmnet for larger numbers of removed nodes.

Acknowledgements. This research was supported in part by the Netherlands Organization for Scientific Research (NWO) with project number 439.16.107, the National Research Foundation (NRF), Prime Minister's Office, Singapore, under its National Cybersecurity R & D Programme (Award No. NRF 2014NCR-NCR001-40) and administered by the National Cybersecurity R & D Directorate, by the Spanish Ministry of Science and Innovation project GIROS TEC2015-66412-R and by the Generalitat de Catalunya research support program SGR-1469.

References

1. Albert, R., Jeong, H., Barabási, A.L.: Error and attack tolerance of complex networks. Nature **406**(6794), 378–382 (2000)
2. Almendral, J.A., Díaz-Guilera, A.: Dynamical and spectral properties of complex networks. New J. Phys. **9**(6), 187 (2007)
3. Barahona, M., Pecora, L.M.: Synchronization in small-world systems. Phys. Rev. Lett. **89**(5), 054101 (2002)
4. Baras, J.S., Hovareshti, P.: Efficient and robust communication topologies for distributed decision making in networked systems. In: Proceedings of the 48th IEEE Conference on Decision and Control, pp. 3751–3756 (2009)
5. Cvetković, D., Simić, S.: Graph spectra in computer science. Linear Algebra Appl. **434**(6), 1545–1562 (2011)
6. Cvetković, D.M.: Applications of graph spectra: an introduction to the literature. Appl. Graph Spectra **13**(21), 7–31 (2009)
7. Donetti, L., Hurtado, P.I., Munoz, M.A.: Entangled networks, synchronization, and optimal network topology. Phys. Rev. Lett. **95**(18), 188701 (2005)
8. Ellens, W., Spieksma, F., Van Mieghem, P., Jamakovic, A., Kooij, R.E.: Effective graph resistance. Linear Algebra. Appl. **435**(10), 2491–2506 (2011)
9. Ellens, W., Kooij, R.E.: Graph measures and network robustness. arXiv preprint arXiv:1311.5064 (2013)
10. Estrada, E.: Characterization of 3D molecular structure. Chem. Phys. Lett. **319**(5), 713–718 (2000)
11. Estrada, E.: When local and global clustering of networks diverge. Linear Algebra Appl. **488**, 249–263 (2016)
12. Estrada, E., Rodriguez-Velazquez, J.A.: Subgraph centrality in complex networks. Phys. Rev. E **71**(5), 056103 (2005)
13. Fiedler, M.: Algebraic connectivity of graphs. Czech. Math. J. **23**(2), 298–305 (1973)
14. Hines, P., Balasubramaniam, K., Sanchez, E.C.: Cascading failures in power grids. IEEE Potentials **28**(5), 24–30 (2009)
15. Jamakovic, A., Van Mieghem, P.: On the robustness of complex networks by using the algebraic connectivity. In: Das, A., Pung, H.K., Lee, F.B.S., Wong, L.W.C. (eds.) NETWORKING 2008. LNCS, vol. 4982, pp. 183–194. Springer, Heidelberg (2008). https://doi.org/10.1007/978-3-540-79549-0_16
16. Jun, W., Barahona, M., Yue-Jin, T., Hong-Zhong, D.: Natural connectivity of complex networks. Chin. Phys. Lett. **27**(7), 078902 (2010)
17. Karrer, B., Levina, E., Newman, M.E.J.: Robustness of community structure in networks. Phys. Rev. E **77**(4), 046119 (2008)
18. Knight, S., Nguyen, H.X., Falkner, N., Bowden, R., Roughan, M.: The Internet topology zoo. IEEE J. Sel. Areas Commun. **29**(9), 1765–1775 (2011)
19. Li, C., Wang, H., De Haan, W., Stam, C.J., Van Mieghem, P.: The correlation of metrics in complex networks with applications in functional brain networks. J. Stat. Mech. Theory Exp. **25**(11), P11018 (2011)
20. Li, T., Fu, M., Xie, L., Zhang, J.F.: Distributed consensus with limited communication data rate. IEEE Trans. Autom. Control **56**(2), 279–292 (2011)
21. Manzano, M., Sahneh, F.D., Scoglio, C.M., Calle, E., Marzo, J.L.: Robustness surfaces of complex networks. Nature Sci. Rep. **4**(6133), 1–6 (2014)
22. Marcus, C.M., Westervelt, R.M.: Stability of analog neural networks with delay. Phys. Rev. A **39**(1), 347 (1989)

136   X. Wang et al.

23. Marzo, J.L., Calle, E., Gomez-Cosgaya, S., Rueda, D., Manosa, A.: On selecting the relevant metrics of network robustness. In: 10th International Workshop on Reliable Networks Design and Modeling (RNDM) (2018)
24. McKay, B.D., Piperno, A.: Practical graph isomorphism, II. J. Symbolic Comput. **60**, 94–112 (2014)
25. Strogatz, S.H.: From Kuramoto to Crawford: exploring the onset of synchronization in populations of coupled oscillators. Phys. D Nonlinear Phenom. **143**(1), 1–20 (2000)
26. Trajanovski, S., Martín-Hernández, J., Winterbach, W., Van Mieghem, P.: Robustness envelopes of networks. J. Complex Netw. **1**(1), 44–62 (2013)
27. Van Mieghem, P.: Graph Spectra for Complex Networks. Cambridge University Press, Cambridge (2010)
28. Van Mieghem, P., Omic, J., Kooij, R.E.: Virus spread in networks. IEEE/ACM Trans. Netw. **17**(1), 1–14 (2009)
29. Wang, X., Koç, Y., Derrible, S., Ahmad, S.N., Pino, W.J., Kooij, R.E.: Multi-criteria robustness analysis of metro networks. Phys. A Stat. Mech. Appl. **474**, 19–31 (2017)
30. Wang, X., Koç, Y., Kooij, R.E., Van Mieghem, P.: A network approach for power grid robustness against cascading failures. In: 7th International Workshop on Reliable Networks Design and Modeling (RNDM), pp. 208–214. IEEE (2015)
31. Wang, X., Pournaras, E., Kooij, R.E., Van Mieghem, P.: Improving robustness of complex networks via the effective graph resistance. Eur. Phys. J. B **87**(9), 1–12 (2014)
32. Watanabe, T., Masuda, N.: Enhancing the spectral gap of networks by node removal. Phys. Rev. E **82**(4), 046102 (2010)
33. Wu, J., Barahona, M., Tan, Y.J., Deng, H.Z.: Spectral measure of structural robustness in complex networks. IEEE Trans. Syst. Man Cybern.-Part A Syst. Hum. **41**(6), 1244–1252 (2011)
34. Wu, Z.X., Holme, P.: Onion structure and network robustness. Phys. Rev. E **84**(2), 026106 (2011)
35. Zanin, M., et al.: Combining complex networks and data mining: why and how. Phys. Rep. **635**, 1–44 (2016)
36. Zeng, Y., Liang, Y.C.: Eigenvalue-based spectrum sensing algorithms for cognitive radio. IEEE Trans. Commun. **57**(6), 1784–1793 (2009)

QoS Criteria for Energy-Aware Switching Networks

Mariusz Głąbowski[1]([✉])(ID), Maciej Stasiak[1], and Michał D. Stasiak[2]

[1] Faculty of Electronics and Telecommunications, Poznan University of Technology,
Polanka 3, 60-965 Poznan, Poland
`mariusz.glabowski@put.poznan.pl`
[2] Department of Investment and Real Estate,
Poznan University of Economic and Business,
Al. Niepodległości 10, 60-875 Poznan, Poland
`michal.stasiak@ue.poznan.pl`

Abstract. This article proposes a method to determine the QoS parameters for energy-aware multiservice switching networks. The initial assumption is that a decrease in the power uptake by the network can be achieved by a temporary switch-off of a certain number of switches. To this end, the article develops methods for a determination of the blocking probability in switching networks with a variable number of switches. The results of the analytical calculations are then compared with the results of simulation experiments for a selected number of structures of switching networks. The study reveals the good accuracy of the proposed model. The results obtained in the study can be applied in constructing energy-aware switching networks.

Keywords: Switching network · Multiservice traffic ·
Energy-aware systems

1 Introduction

Works on network energy demand have been carried out at a number of levels: from the lowest one, involving the construction of devices, through the improvement of algorithms that control the operation of individual interfaces, up to the highest level related to dimensioning and designing of networks and systems. It is plain to see that the power consumption of network devices largely depends on the volume of offered traffic [2,28]. While choosing appropriate devices, network operators target economic efficiency and service quality and rely on and are guided by the maximum traffic value in those periods that are deemed to provide the heaviest load for systems. This, in turn, leads to the application of "over-dimensioned" devices throughout most of the 24-hour working time. In such circumstances, a temporary switch-off of certain elements in network devices or, should the need arise some of their modules, seems to be an appropriate solution to the problem [15].

T. Q. Duong et al. (Eds.): Qshine 2018, LNICST 272, pp. 137–147, 2019.
https://doi.org/10.1007/978-3-030-14413-5_11

Works on optimization of the operation of network nodes in relation to the minimization criterion for power consumption have been conducted in a large number of company research centers and academic institutions, e.g. [5,12,15,18,27–29]. One of the possibilities to minimize energy uptake is to apply appropriate traffic engineering algorithms [2,15,28]. More and more often, connections between servers in data centers have topologies that correspond to the structures of switching networks (SN). Within this particular context, an application of appropriately selected algorithms can lead to a decrease in the demand for energy in data centers [5,12,18].

This article discusses the emerging possibilities that can be exploited to design optimum energy-aware SNs. A construction of such networks is based on the two following criteria: minimization of power uptake and minimization of the blocking probability. The former criterion can be satisfied by utilization of appropriate mechanisms for deactivation of individual elements of a SN during network low-load periods. This means that a certain number of the elements of a SN (e.g. switches) will be temporarily "removed" from the network structure. However, the two above mentioned criteria contradict each other – switching off of certain elements causes the blocking probability to be increased. Therefore, the optimization process should be intertwined with a certain assumed and pre-defined boundary level for the Quality of Service (QoS) parameters.

The SN analysis is based on effective availability models. This concept is proposed for two-stage single-service SNs in [1,3] and then generalized to include any number of stages [6,17,25]. A number of effective availability models are also proposed for multiservice SNs [8,23,26]. The literature of the subject also offers models that expand the range of possible applications of effective availability methods, e.g. [7,9–11,13,24].

This article discusses a possibility of a determination of boundary QoS parameters for energy-aware SNs. The influence of changes in the SN structure on the blocking probability of individual traffic classes is then determined. For this purpose, an appropriately modified Point to Group Blocking for Multichannel Traffic (PGBMT) method for a SN operating in the point-to-group selection mode [23,26] is used. In the proposed model, the blocking probability for successive SN structures in which the number of switches is decreased is determined. The obtained results can be applied to construct energy-aware SNs structures that would satisfy the adopted QoS assumptions. The present article is structured as follows. Section 2 provides a description of a Clos multiservice SN. Section 3 includes a discussion of an analytical model of a multiservice switching network in which the number of switches of the middle stage is variable. In Sect. 4, the results of the analytical calculations are compared with the results of the simulation experiments. Section 5 sums up the article.

2   Multiservice Switching Network

Figure 1 shows the structure of a three-stage Clos SN [4,30]. Each stage of the SN has k symmetrical switches with $k \times k$ links. All links have the capacity equal

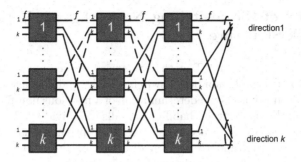

Fig. 1. Three-stage Clos switching network

to f Allocation Units (AUs). A single AU is expressed in kbps and is defined as the Greatest Common Divisor (GCD) of bitrates of all traffic classes [19,21]. The output links of the SN form the so-called output directions. The assumption is that the i-th output links of each of the switches of the last stage create an i-th direction.

If a switch of the middle stage, for example the first switch in Fig. 1, is removed from the network, then all links that connect this switch with the switches of the neighboring (adjacent) switches (marked by the dotted line in Fig. 1) will also be removed. Therefore, if we remove u switches of the middle stage from the network structure, we will obtain a network comprised of k switches of the type $k \times (k-u)$ in the first stage, $(k-u)$ switches of the type $k \times k$ in the second stage and k switches of the type $(k-u) \times k$ in the third stage. The initial assumption is that traffic offered to the SN is a mixture M of Erlang traffic streams that can be characterized by the following parameters: A_i – the average intensity of traffic of class i ($1 \leq i \leq M$) offered to links of the switching network, t_i – the number of AUs ($1 \leq t_i \leq f$) required for a connection of class i in the SN to be set up.

A further assumption is that the SN operates in the point-to-group selection mode. The control algorithm determines first a switch of the first stage at the input of which a new call has arrived. Then, the algorithm determines the switches of the last stage that have free outputs in the demanded direction. If all links of a given direction are occupied (busy), then the phenomenon of internal blocking will occur and the call will be discarded and lost. If even one link of a given direction is free, then the algorithm will attempt to set up a connection between the switch of the first stage that has been determined earlier and a switch of the last stage that has a free output in the demanded direction. If the execution of the connection is not possible, the algorithm will check successively a possibility of setting up a connection with other switches of the last stage that have free links in the given direction. If setting up of a connection is still impossible, then the call will be lost due to the phenomenon of internal blocking. Note that in the case of multiservice switching networks, the notion of a free link for calls of class i means that this link has at least t_i free (unoccupied) AUs.

3   A Model of the Switching Network with Variable Network Structure

The modified version of PGBMT method was used in the paper to model multi-service SNs with a variable number of switches in the middle stage [23]. The idea behind the method is based on a determination of a fictitious non-full-availability system in which the blocking probability for calls of individual classes is exactly the same as in the SN. The accompanying assumption is that traffic offered to the fictitious system and to a given direction of the network is identical. The fictitious system is indicated by the availability parameter. In the PGBMT method, the value of the availability for particular traffic classes, called effective availability, can be determined on the basis of the structure and load of the SN. The method is composed of the following elements: a model of inter-stage links, model of output links and a method to determine effective availability for each of traffic classes. These three elements enable us to determine the blocking probability for call streams that are offered in the SN. In this section, a method for a determination of effective availability for a variable number of switches of the middle stage is proposed.

3.1   Model of Inter-stage Links

A full-availability group (FAG) with multiservice traffic [14,20] can serve as a model of inter-stage links. The group has the capacity f AUs. The occupancy distribution $[P_n]_f$ and the blocking probability $E_{i,\mathrm{FAG}}$ for calls of class i in a link with the capacity f AUs can be described as follows:

$$n\,[P_n]_f = \sum_{i=1}^{M} A_{i,\mathrm{FAG}} t_i\,[P_{n-t_i}]_f\,, \tag{1}$$

$$E_{i,\mathrm{FAG}} = \sum_{n=f-t_i+1}^{f} [P_n]_f\,, \tag{2}$$

where n is the number of busy AUs in the FAG with the capacity f AUs and $A_{i,\mathrm{FAG}}$ is the intensity of traffic of class i offered to the FAG. If A_i is the intensity of traffic of class i, whereas u is the number of removed switches of the middle stage, then in symmetry strength of the SN from Fig. 1 we can write:

$$A_{i,\mathrm{FAG}} = A_i/\,[k(k-u)]. \tag{3}$$

3.2   Model of Output Links

The output group (direction) of the SN from Fig. 1 comprises of k links with the capacity f AUs. To model the direction, a limited-availability group (LAG) can be used [22]. The group is composed of k separated links. The concept of separation results from the underlined method of service. A call of class i can be serviced only when the LAG has at least t_i free AUs in a single link. This means that a call cannot be "divided" between AUs of a number of links. This method for service corresponds to the operation of SNs, where each call is always serviced by only one link in a given direction. The occupancy distribution $[P_n]_V$ in a LAG with the capacity $V = kf$ can be then expressed by the following formula:

$$n\,[P_n]_V = \sum_{i=1}^{M} A_{i,\text{LAG}} t_i \omega_i(n - t_i)\,[P_{n-t_i}]_V\,, \tag{4}$$

where $A_{i,\text{LAG}}$ is the intensity of traffic of class i offered to a given direction:

$$A_{i,\text{LAG}} = A_i / k. \tag{5}$$

The parameter $\omega_i(n)$ in (4) is the conditional transition probability between states n and $(n + t_i)$. The parameter determines the probability of such a distribution of free AUs in the LAG that makes service of a call of class i in state n possible:

$$\omega_i(n) = [F(V - n, k, f, 0) - F(V - n, k, t_i - 1, 0)]\,/\,[F(V - n, k, f, 0)]. \tag{6}$$

The combinatorial function $F(x, k, f, h)$ determines the number of arrangements of x elements (free AUs) in k sets (LAG links), each with the capacity f elements (AUs). The assumption is that in each set h elements have been earlier deployed (accommodated):

$$F(x, k, f, h) = \sum_{g=0}^{\lfloor \frac{x - kh}{f - t + 1} \rfloor} (-1)^g \binom{k}{g} \binom{x - k(h-1) - 1 - g(f - h + 1)}{k - 1}. \tag{7}$$

On the basis of (4), the distribution of free links $[P_s(i)]_k$ can be derived. This distribution determines the probability that s links can serve a call of class i [22]:

$$[P_s(i)]_k = \sum_{n=0}^{V=kf} \frac{\binom{k}{s} \sum_{z=st_i}^{\Psi} F(z, s, f, t_i) F(V - n - z\,k - s, t_i - 1, 0)}{F(V - n, k, f, 0)} [P_n]_V\,, \tag{8}$$

where $\Psi = sf$ for $n \le V - sf$ and $\Psi = V - n$ for $n > V - sf$. Note that for $s = 0$, the probability $[P_0(i)]_k$ determines the blocking probability for calls of class i in the LAG.

3.3 Blocking Probability in Switching Networks

In the PGBMT method, the internal blocking probability $[E_{\text{int}}(i)]_{\text{SN}}$ for traffic of class i in the multiservice SN is defined by the following formula:

$$[E_{\text{int}}(i)]_{\text{SN}} = \sum_{s=1}^{k-d(i)} \frac{[P_s(i)]_k}{1 - [P_0(i)]_k} \left\{ \binom{k-s}{d(i)} \middle/ \binom{k}{d(i)} \right\}, \tag{9}$$

where $d(i)$ is the effective availability for calls of class i (Sect. 3.4). Formula (9) determines the probability that all $d(i)$ available switches of the last stage have occupied links in a given direction. In (9), a truncated distribution $[P_s(i)]_k/(1-[P_0(i)]_k)$ of free links is used because the event of internal blocking occurs exclusively in the case of the existence of at least one free link in a given direction. Therefore, a situation in which all links in a given direction are occupied must be excluded. To determine the external blocking probability $[E_{\text{ext}(i)}]_{\text{SN}}$ for calls of class i the distribution $[P_s(i)]_k$ is used:

$$[E_{\text{ext}}(i)]_{\text{SN}} = [P_0(i)]_k . \tag{10}$$

The control algorithm first checks the existence of free links in a given direction, i.e. checks whether the phenomenon of external blocking occurs or not. If not, then the algorithm attempts to set up a connecting path in the SN. Taking, therefore, into account the operation of the algorithm, the total blocking probability in the SN is the sum of the external blocking probability and the internal blocking probability, with the exclusion of a possibility of a concurrent occurrence of the event of external and internal blocking:

$$[E(i)]_{\text{SN}} = [E_{\text{ext}}(i)]_{\text{SN}} + [E_{\text{int}}(i)]_{\text{SN}} \cdot [1 - [E_{\text{ext}}(i)]_{\text{SN}}] . \tag{11}$$

3.4   Effective Availability for Calls of Class i

The PGBMT method is based on reducing a multi-stage SN to a fictitious single-stage system, the so-called non-full availability group, characterized by the effective availability parameter. The parameter is defined as such an availability to switches of the last stage for which blocking probabilities in the SN and the non-full availability are identical. Effective availability for a given call class can be determined on the basis of the so-called equivalent network (EN) that has the same structure as the SN, but the capacity of the links equal to 1AU. While making a determination of the effective availability for a given traffic class the assumption is then that the EN services only this class and the load of each of the links in the EN is equal to the blocking probability of a given class of calls in a SN link. Effective availability for traffic of class i in a three-stage SN can be determined by the following formula [23]:

$$d_u(i) = [1 - \pi_u(i)] k + \pi_u(i)b(i) \{1 + [k - a_u(i)] [1 - a_u(i)]\}, \tag{12}$$

where:

- $\pi_u(i)$ – the probability of unavailability of one switch of the third stage in SN in which u switches of the middle stage have been removed,
- $a_u(i)$ – fictitious load of EN link for calls of class i, determined on the basis of the FAG model (Formulas (1)–(3)):

$$a_u(i) = E_{i,\text{FAG}}, \tag{13}$$

- $b(i)$ – fictitious load at the EN output. This parameter is determined by Formulas (1)–(3), with taking into account the fact that the number of links in a given direction is fixed (constant) and does not depend on removed switches of the middle stage taken into consideration. Hence, in (3) the assumption should be that $u = 0$.

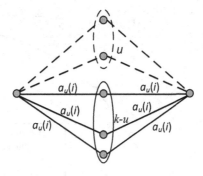

Fig. 2. Channel graph of the three-stage switching network from Fig. 1

The parameter $\pi_u(i)$ can be determined on the basis of an EN channel graph that shows all possible connecting paths between the switches of the external stages. Figure 2 shows the SN channel graph from Fig. 1. Each of its edges is attributed the fictitious load $a_u(i)$. The probability $\pi_u(i)$ corresponds to the blocking of all connecting paths and can be determined on the basis of the method [16]. For the graph from Fig. 2, in which u switches of the middle stage have been removed, we then have:

$$\pi_u(i) = \left\{ 1 - [1 - a_u(i)]^2 \right\}^{k-u}. \tag{14}$$

In Formula (14), the expression $1 - a_u(i)$ means that the graph edge is free, therefore $1 - [1 - a_u(i)]^2$ determines an event of blocking of a connecting path. By raising this expression to the power $(k - u)$ we can determine the blocking events for all connecting paths.

3.5 Commentary

A possibility of changing the SN structure, which consists in a temporary deactivation of a certain number of switches, is assumed in works on energy-aware SNs. This process is correlative with a pre-defined boundary level of blocking. This means that in any circumstances and under any conditions the SN structure can be changed in such a way as to prevent the blocking probability from exceeding the required values. The proposed modification to the PGBMT method maps any changes in the SN structure into changes in the value of the effective availability parameter. As a result, the method allows boundary blocking probabilities for particular call classes for a variable number of switches of the middle stage of a three-stage Clos SN to be evaluated.

4 Numerical Examples

The proposed model of a multiservice SN with a variable number of switches of the middle stage is an approximate model. To determine its applicability for

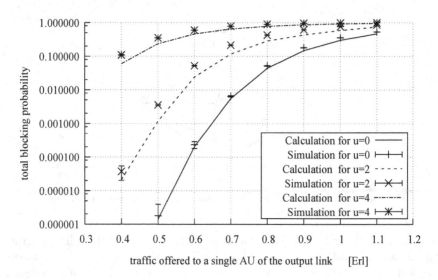

Fig. 3. Blocking probability for class 3 calls ($t_3 = 6$ AUs) in the switching networks with variable number of active switches in the middle stage

modeling energy-aware networks, the results of the analytical calculations were compared with the results of the simulations for a selected number of Clos SNs (Fig. 1). A network with the following parameters was chosen for modeling:

- size of the switch of the first stage: $8 \times (8 - u)$,
- size of the switch of the second stage: 8×8,
- size of the switch of the third stage: $(8 - u) \times 8$,
- capacity of the input, output and inter-stage link $f = 30$ AUs,
- the number of classes of offered traffic $M = 3$,
- the demanded number of AUs for offered traffic classes: $t_1 = 1$ AU, $t_2 = 2$ AUs, $t_3 = 6$ AUs,
- proportions of the offered traffic mixture: $A_1 t_1 : A_2 t_2 : A_3 t_3 = 1 : 1 : 1$.

Figure 3 shows the results of the modeling of the SN under consideration for the number of switches of the middle stage equal to 8 ($u = 0$), 6 ($u = 2$) and 4 ($u = 4$). Due to the limited length of the paper, the results for the class with the maximum demands are only presented. The results of the simulations are shown in the form of points with 95 % confidence intervals determined on the basis of the t-Student distribution for 10 series with 10000000 calls of each class in each series. The results of the analytical modeling are presented in the form of lines.

The results obtained in the study reveal good accuracy of the proposed SN model with a variable number of switches in the middle stage. All the results are expressed in relation to the value of traffic offered to one AU of the output link in SN:

$$a_{\text{out}} = \sum_{i=1}^{M} [A_i t_i] / k^2. \tag{15}$$

Let us assume that the acceptable value of the blocking probability is 5%. While analyzing the results presented in Fig. 3 one can observe that with a decrease in the value a_{out} from 0.8 Erl./AU to the level of 0.6 Erl./AU it is feasible to use a SN with 6 switches in the middle stage, instead of 8 switches, in order to keep the assumed value of blocking.

5   Conclusions

This article proposes a new method to model multiservice SNs with a variable structure that allows the number of switches of the middle stage of an SN to be decreased. The application of a variable load-dependent structure makes it possible to construct energy-aware networks. The present article does not discuss the issue of constructing such SNs. What the article is focused on is to show how and in what way a change in the structure of a SN can influence the blocking probability for individual traffic classes. As a result, the method proposed in the article can be used in practice to help solve designing and optimization issues with regard to constructing energy saving algorithms by way of limiting the number of active elements to values that secure acceptable level of the QoS parameters.

The intention of the authors is to pursue the work initiated in the article and to develop, on the basis of the proposed method, appropriate algorithms for temporary activation/deactivation of a certain number of switches, which in consequence will allow a lowered power consumption by a switching network to be quantitatively estimated. In the further works, in order to increase the accuracy of the model, blocking probability phenomenon of the first stage switches will be incorporated to the model.

Acknowledgements. This paper was developed as a result of the research project 2016/23/B/ST7/03925 entitled "Modelling and service quality evaluation of Internet-based services" funded by the National Science Centre.

References

1. Binida, N., Wend, W.: Die effektive erreichbarkeit für abnehmerbundel hinter zwischenleitungsanungen. Nachrichtentechnische Zeitung (NTZ) **11**(12), 579–585 (1959)
2. Chabarek, J., Sommers, J., Barford, P., Estan, C., Tsiang, D., Wright, S.: Power awareness in network design and routing. In: IEEE INFOCOM 2008 - the 27th Conference on Computer Communications, April 2008. https://doi.org/10.1109/INFOCOM.2008.93
3. Charkiewicz, A.: An approximate method for calculating the number of junctions in a crossbar system exchange. Elektrosvyaz **2**, 55–63 (1959)
4. Clos, C.: A study of non-blocking switching networks. Bell Syst. Tech. J. **32**, 406–424 (1953)
5. Cordeschi, N., Shojafar, M., Baccarelli, E.: Energy-saving self-configuring networked data centers. Comput. Netw. **57**(17), 3479–3491 (2013). https://doi.org/10.1016/j.comnet.2013.08.002

6. Ershova, E., Ershov, V.: Digital Systems for Information Distribution. Radio and Communications, Moscow (1983). (in Russian)
7. Głąbowski, M., Sobieraj, M.: Analytical modelling of multiservice switching networks with multiservice sources and resource management mechanisms. Telecommun. Syst. **66**(3), 559–578 (2017). https://doi.org/10.1007/s11235-017-0305-4
8. Głąbowski, M.: Recurrent method for blocking probability calculation in multiservice switching networks with BPP traffic. In: Thomas, N., Juiz, C. (eds.) EPEW 2008. LNCS, vol. 5261, pp. 152–167. Springer, Heidelberg (2008). https://doi.org/10.1007/978-3-540-87412-6_12
9. Głąbowski, M., Stasiak, M.: Point-to-point blocking probability in switching networks with reservation. Ann. Telecommun. **57**(7–8), 798–831 (2002)
10. Głąbowski, M., Stasiak, M.D.: Modelling of multiservice switching networks with overflow links for any traffic class. IET Circuits Devices Syst. **8**(5), 358–366 (2014). https://doi.org/10.1049/iet-cds.2013.0430
11. Głąbowski, M., Stasiak, M.D.: Multiservice switching networks with overflow links and resource reservation. Math. Prob. Eng. **2016**, 17 (2016). https://doi.org/10.1155/2016/4090656. Article ID 4090656
12. Gyarmati, L., Trinh, T.A.: How can architecture help to reduce energy consumption in data center networking? In: Proceedings of 1st International Conference on Energy-Efficient Computing and Networking, e-Energy 2010, pp. 183–186. ACM, New York (2010). https://doi.org/10.1145/1791314.1791343
13. Hanczewski, S., Sobieraj, M., Stasiak, M.D.: The direct method of effective availability for switching networks with multi-service traffic. IEICE Trans. Commun. **E99–B**(6), 1291–1301 (2016)
14. Kaufman, J.: Blocking in a shared resource environment. IEEE Trans. Commun. **29**(10), 1474–1481 (1981)
15. Kühn, P.J.: Systematic classification of self-adapting algorithms for power-saving operation modes of ICT systems. In: Proceedings of 2nd International Conference on Energy-Efficient Computing and Networking, pp. 51–54. ACM, New York (2011). https://doi.org/10.1145/2318716.2318724
16. Lee, C.: Analysis of switching networks. Bell Syst. Tech. J. **34**(6), 1287–1315 (1955)
17. Lotze, A., Roder, A., Thierer, G.: PCM-charts. Technical report, Institute of switching and data technics, University of Stuttgard (1979)
18. Niewiadomska-Szynkiewicz, E., Sikora, A., Arabas, P., Kamola, M., Mincer, M., Kołodziej, J.: Dynamic power management in energy-aware computer networks and data intensive computing systems. Future Gener. Comput. Syst. **37**, 284–296 (2014). https://doi.org/10.1016/j.future.2013.10.002
19. Pras, A., Nieuwenhuis, L., van de Meent, R., Mandjes, M.: Dimensioning network links: a new look at equivalent bandwidth. IEEE Netw. **23**(2), 5–10 (2009). https://doi.org/10.1109/MNET.2009.4804330
20. Roberts, J.: A service system with heterogeneous user requirements - application to multi-service telecommunications systems. In: Pujolle, G. (ed.) Proceedings of Performance of Data Communications Systems and their Applications, pp. 423–431. North Holland, Amsterdam (1981)
21. Roberts, J. (ed.): Performance evaluation and design of multiservice networks. Final Report COST 224. Commission of the European Communities, Brussels (1992)
22. Stasiak, M.: Blocking probability in a limited-availability group carrying mixture of different multichannel traffic streams. Ann. Télécommun. **48**(1–2), 71–76 (1993)
23. Stasiak, M.: Combinatorial considerations for switching systems carrying multichannel traffic streams. Ann. Télécommun. **51**(11–12), 611–625 (1996)

24. Stasiak, M., Zwierzykowski, P.: Point-to-group blocking in the switching networks with unicast and multicast switching. Perform. Eval. **48**(1–4), 249–267 (2002)
25. Stasiak, M.: Blocage interne point a point dans les reseaux de connexion. Ann. Télécommun. **43**(9–10), 561–575 (1988)
26. Stasiak, M., Głąbowski, M.: Multi-service switching networks with point-to-group selection and several attempts of setting up a connection. In: Kouvatsos, D. (ed.) Performance Modelling and Analysis for Heterogeneous Networks, pp. 3–26. River Publishers, Aalborg (2009)
27. Tucker, R.S., Parthiban, R., Baliga, J., Hinton, K., Ayre, R.W.A., Sorin, W.: Evolution of WDM optical IP networks: a cost and energy perspective. J. Lightwave Technol. **27**(3), 243–252 (2009). http://jlt.osa.org/abstract.cfm?URI=jlt-27-3-243
28. Vasić, N., Kostić, D.: Energy-aware traffic engineering. In: Proceedings of 1st International Conference on Energy-Efficient Computing and Networking, e-Energy 2010, pp. 169–178. ACM, New York (2010). https://doi.org/10.1145/1791314.1791341
29. Venkatachalam, V., Franz, M.: Power reduction techniques for microprocessor systems. ACM Comput. Surv. **37**(3), 195–237 (2005). https://doi.org/10.1145/1108956.1108957
30. Żal, M., Wojtysiak, P.: An energy-efficient control algorithms for switching fabrics. In: 2014 16th International Telecommunications Network Strategy and Planning Symposium (Networks), pp. 1–5, September 2014. https://doi.org/10.1109/NETWKS.2014.6959228

Modelling Overflow Systems with Queuing in Primary Resources

Mariusz Głąbowski$^{(\boxtimes)}$, Damian Kmiecik, and Maciej Stasiak

Faculty of Electronics and Telecommunications, Poznan University of Technology,
Poznań, Poland
`mariusz.glabowski@put.poznan.pl`

Abstract. This article proposes a new method to determine the characteristics of multiservice overflow systems with queueing systems. A number of methods have been developed that have the advantage of determining the parameters of traffic directed to secondary resources as well as providing a way to model these resources. The accompanying assumption is that queues with limited capacities are used in primary resources. The results of analytical calculations are compared with the results of simulation experiments for a number of selected structures of overflow systems with queueing in primary resources. The results of the study confirm high accuracy of the proposed method and, in consequence, the accuracy of the theoretical assumptions adopted for the proposed method.

Keywords: Overflow system with queueing · Multiservice traffic ·
Blocking probability

1 Introduction

Traffic overflow is one of the oldest and best known mechanisms for traffic distribution optimization in networks. Traffic overflow is based on the principle that when certain resources, called primary resources, are fully occupied, the traffic overflow mechanism allows calls that are still offered to the resources to be directed (i.e. to overflow) to some alternative resources, called secondary resources [24].

Overflow systems in single-service telecommunications systems with losses have been widely addressed in the literature, e.g. [5,26]. In [26], the analytical modeling of single-service overflow systems is expanded to include the ERT method (Equivalent Random Traffic). The work [5] proposes a simple method for an analysis of overflow systems, i.e. the so-called Hayward's method based on a modification of Erlang B Formula. The author of [13] points out the fact that the call stream offered to secondary resources can be approximated, with acceptable accuracy, by a Pascal call stream. Single-service overflow models with queueing have been addressed is a large number of works [15,18–20]. The queue to which overflow traffic with the peakedness factor higher than unity is offered is analyzed

T. Q. Duong et al. (Eds.): Qshine 2018, LNICST 272, pp. 148–157, 2019.
https://doi.org/10.1007/978-3-030-14413-5_12

in [11]. Works such as [15,20] are devoted to the analysis of the two-dimensional Markov process in an overflow system in which queues are introduced both to primary and secondary resources. An overflow system that services two traffic streams, one of which is queued is described in [3]. In [18] and [19], overflow traffic in systems with queues is described on the basis of the two and three first moments of the IPP process (Interrupted Poisson Process) [16].

The problem of multiservice traffic overflow has been widely discussed in the literature. The article [17] analyzes overflow systems on the basis of Markov-Modulated Poisson Process (MMPP), while in [14] these systems are analysed on the basis of Batched Poisson Process (BPP). The works [4,14] provide analyses of traffic that is characterized by an appropriate peakedness factor that can approximate overflow traffic. A methodology for dimensioning multiservice overflow systems is proposed in [7,12]. In the proposed methods, Hayward's approach is generalized [5] to model secondary resources, which is based on a division of the averaged values offered to secondary resources and the capacities of secondary resources by the peakedness factors of relevant traffic scenarios. The paper [6] considers a model in which secondary resources are of distributed nature, which means that they are composed of a number of separated resources with full availability [8]. In [25], overflow systems in which traffic can change the service parameters, such as the service time and bitrate in secondary resources, are analyzed. Then, in [6], a possibility of elastic traffic service in the overflow system is investigated. As yet, multiservice overflow models with queueing have not been considered in the literature.

This work proposes a model of a multiservice overflow system with queueing for primary resources. To model a queueing system of primary resources, a multiservice queueing model with state-dependent service disciple called SD FIFO (State Dependent FIFO) is used [10,23]. The model is based on a queueing interpretation of the system of resources with losses that supports elastic traffic [21,22]. The SD FIFO discipline corresponds to a resource distribution in a multiservice server according to the balanced fairness algorithm [1,9] that approximates very well other resource distribution algorithms, e.g. the proportional (with regard to offered traffic) algorithm. To model secondary resources, this article takes advantage of the generalized Hayward model [7,12]. The article is structured as follows. Section 2 provides an outline of the subject of research, i.e. a multiservice overflow system in which calls offered to primary resources can be queued. Section 3 includes a model of primary resources, description of a method for a determination of the parameters of traffic that overflows from these resources and a model of secondary resources. In Sect. 4, the results of analytical calculations are compared with the results of simulation experiments for a number of selected structures of overflow systems with queueing for calls in primary (and secondary) resources. Section 5 sums up the article.

2 Traffic Overflow Systems with Queueing for Primary Resources

This study considers multiservice overflow systems in which, when primary resources are fully occupied, calls of different traffic classes will be first directed to queues, where they will stay until enough resources to serve them have been released. Until now, the literature of the subject on multiservice overflow systems has not considered a possibility of queueing of calls before they are directed to secondary resources. The assumption in this article is that queues have a limited capacity and that calls that cannot wait in a queue will be directed to a system of secondary resources. Figure 1 shows a general diagram of the traffic overflow system with queueing in primary resources. The following notation is used:

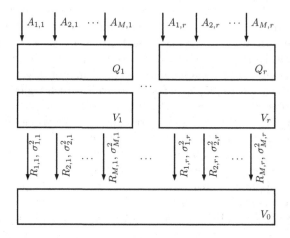

Fig. 1. Diagram of a multiservice overflow system with queues in primary resources

- r – number of component primary resources in the overflow system,
- Q_k – queue capacity in primary resource k $(1 \leq k \leq r)$,
- V_k – capacity of primary resource k $(1 \leq k \leq r)$,
- V_0 – capacity of secondary resource,
- M – number of traffic classes of traffic offered in overflow system,
- $A_{i,k}$ – traffic intensity of class i $(1 \leq i \leq M)$ offered to resources k $(1 \leq k \leq r)$,
- $R_{i,k}$ – traffic intensity of class i $(1 \leq i \leq M)$ that overflows from resource k $(1 \leq k \leq r)$,
- $\sigma^2_{i,k}$ – variance of traffic intensity of class i $(1 \leq i \leq M)$ that overflows from resource k $(1 \leq k \leq r)$.

2.1 Traffic Offered to Primary Resources

Modern networks, including the Internet, are packet networks. Packets that belong to a service that is being executed create traffic streams that can be analysed in exactly the same way as calls (or flows) are analyzed [1,2,21]. A mathematical analysis of the internal structure of these streams (calls) is very complex and frequently leads to solutions that exclude them from being applied in practice. Hence, multiservice systems are analyzed at the call level. The result of many studies show that calls can be approximated by streams of "Poisson-like" character [1,2]. Such an approach makes it possible to discretize the system and then to construct models that are based on multi-dimensional Markov processes. Discretization is a change in the variable bitrate (VBR) of a packet stream that belongs to a given call in such a way as to convert it into a fixed, constant fictitious bitrate (CBR). The assumption in the most recent works, in particular those that refer to modeling of the TCP/IP network, is that such a fictitious bitrate is equal to the maximum bitrate (flow rate) of a real packet stream that corresponds to calls. The knowledge of fictitious bitrates for individual call classes makes it possible to determine the allocation unit (AUs) of a given overflow system. The maximum value of the AU is described as the Greatest Common Divisor (GCD) of all fictitious bitrates of the calls offered to an overflow system:

$$c_{AU} = \text{GCD}(c_1, c_2, \dots, c_M), \tag{1}$$

where c_i is the maximum bitrate of a packet (call) stream of class i, whereas c_{AU} is the bitrate of the allocation unit. Having determined the value of the AU, both the capacity of the system V and the volume of the resources necessary for a connection of class i to be set up is discretized, i.e. expressed in AUs:

$$V = \frac{C}{c_{AU}}, \tag{2}$$

$$t_i = \frac{c_i}{c_{AU}}, \tag{3}$$

where C the total bitrate of a given resource of the system. The assumption in the model proposed in this article is that both the capacity of the primary and secondary resources and the demanded volume of resources necessary for calls of individual classes to be set up is expressed in AUs.

2.2   Primary and Secondary Resources

Fig. 1 shows a general system for traffic overflow. The primary resources are composed of r component primary resources, where each resource k has the capacity V_k, expressed in AUs. Each component resource can service M Erlang traffic classes. If a call of class i cannot be serviced by a primary resource k, it is directed to a queue in the component resource k with the capacity Q_k AUs. A lack of free space, i.e. t_i AUs, in this queue, will cause the call of class i to be directed to the secondary resource with the capacity of V_0 AUs. A lack of free t_i AUs in the secondary resources will, in turn, lead to irrevocable loss of the call.

3   Models of Resources in the Overflow System

The model of the traffic overflow system with queueing in primary resources is composed of a model of primary resources, a model for a determination of the average value and variance for individual overflow traffic classes and a model of secondary resources. These three models allow us to determine all important characteristics of the overflow system, in particular the blocking probability for call streams offered to the system. To model each of the primary resources, a multiservice SD FIFO model was used [10], whereas to determine the parameters of traffic that overflows from the primary resources, the method developed in [7] was used.

3.1   Model of Component Primary Resources

In the article, to model the component primary resources with queueing, a recurrent occupancy distribution in a queueing SD FIFO system [10] was used that, according to the notation adopted in the article, can be written in the following way:

$$[P(n)]_{V_k+Q_k} = \begin{cases} \frac{1}{n}\sum_{i=1}^{M} A_{i,k} t_i [P(n-t_i)]_{V_k+Q_k} & \text{for } 0 \leq n \leq V_k, \\ \frac{1}{V_k}\sum_{i=1}^{M} A_{i,k} t_i [P(n-t_i)]_{V_k+Q_k} & \text{for } V_k < n \leq V_k + Q_k, \end{cases} \tag{4}$$

where n is the number of AUs occupied by calls that are currently in a component resource k (serviced and queued), whereas the distribution $[P(n)]_{V_k+Q_k}$ defines the occupancy probability n AUs in a system with the capacity of the component resource V_k and the queue capacity Q_k. The blocking probability in such a system results from the finite capacity of the queue and for calls of class i can be determined by the formula:

$$E_{i,k} = \sum_{n=V_k+Q_k-t_i+1}^{V_k+Q_k} [P(n)]_{V_k+Q_k}. \tag{5}$$

Formula (5) determines the sum of all blockable states for calls of class i, i.e. those states in which the number of free AUs in a queue is lower than the number of AUs demanded for a call of class i to be set up. Queuing service discipline for a SD FIFO queue corresponds to a resource allocation for each class of offered traffic on the basis of the balanced fairness algorithm.

The model approximates very well different resource distribution algorithms for serviced calls, in particular the proportional algorithm [10].

3.2   Model of Traffic that Overflows Form Primary Resources

Traffic of class i that overflows from the primary resource k can be described by the three following parameters: the average value of traffic intensity $R_{i,k}$, the variance $\sigma_{i,k}^2$ and the number of AUs t_i necessary for a given connection to be executed. The parameter $R_{i,k}$ can be determined on the basis of the loss in traffic in the component resources with queues that are in the blocking state. Therefore, the average value $R_{i,k}$ of the traffic intensity of traffic of class i that overflows from a component queuing system k, can be, on the basis of (5), determined by the following formula:

$$R_{i,k} = A_{i,k}E_{i,k}. \tag{6}$$

To determine the variance of traffic that overflows from the primary resources, a modification to the method proposed in [12] will be used. The proposed modification is based on a decomposition of each of the component primary resources V_k with the queue Q_k into M fictitious resources with the capacities $V_{i,k}^*$, each of them servicing exclusively calls of one class i with the traffic intensity $A_{i,k}$. The assumption is that the blocking probability of class i in a single-service fictitious resource is exactly the same as the blocking probability of class i in the primary component queueing system k. Since the fictitious resource also services Erlang traffic, then its capacity can be determined on the basis of Erlang's function:

$$E_{i,k} = E_{V_{i,k}^*(A_{i,k})} = \frac{(A_{i,k})^{V_{i,k}^*}}{(V_{i,k}^*)!} \bigg/ \sum_{n=0}^{V_{i,k}^*} \frac{(A_{i,k})^n}{n!}. \tag{7}$$

Now, on the basis of Riordan's formula, the variance can be approximated by the variance of traffic that overflows from the fictitious resources [26]:

$$\sigma_{i,k}^2 = R_{i,k}\left(\frac{A_{i,k}}{V_{i,k}^* + 1 - A_{i,k} + R_{i,k}} + 1 - R_{i,k}\right). \tag{8}$$

The peakedness factor of traffic that overflows from the primary component resources k is then equal to:

$$Z_{i,k} = \frac{\sigma_{i,k}^2}{R_{i,k}}. \tag{9}$$

3.3   Model of Secondary Resources

To model the occupancy distribution in multiservice secondary resources to which a mixture of overflow traffic is offered, the generalized Hayward distribution, proposed in [7] can be used. According to the notation adopted in the article, this distribution can be written in the following way:

$$n\left[P(n)\right]_{V_0/Z_0} = \sum_{i=1}^{M}\sum_{k=1}^{r}\frac{R_{i,k}}{Z_{i,k}}t_i\left[P(n-t_i)\right]_{V_0/Z_0}, \tag{10}$$

where the coefficients $Z_{i,k}$ are determined on the basis of (9), whereas the parameter Z_0 is the so-called aggregate peakedness factor. The factor Z_0 can be determined on the basis of the weighted average of the coefficients $Z_{i,k}$, assuming that the weights are determined by the proportion of AUs demanded by calls of individual classes in relation to the demands of all traffic classes. Therefore:

$$Z_0 = \sum_{i=1}^{M} \sum_{k=1}^{r} Z_{i,k} \frac{R_{i,k} t_i}{\sum_{s=1}^{r} \sum_{j=1}^{M} R_{j,s} t_j}. \tag{11}$$

The blocking probability for traffic of class i in the secondary resources (regardless of the fact from which primary resources this traffic overflows) is equal to:

$$E_{i,0} = \sum_{n=V_0/Z_0-t_i+1}^{V_0/Z_0} [P(n)]_{V_0/Z_0}. \tag{12}$$

Formulas (10)–(12) allow the blocking probabilities in a multiservice overflow system with queues in the primary resources to be determined.

4   Results of Modeling of a Selected Overflow System with Queues in Primary Resources

The method for a determination of the blocking probability in systems with overflow traffic and call queuing capabilities in primary resources presented in the article is an approximate method. To evaluate its accuracy and adopted assumptions for the method, the results of the analytical calculations of the blocking probability in the alternative resources were compared with the data obtained in simulation experiments. Studies were carried out for a large number of network structures with traffic overflow. Due to the space constraints, Figs. 2, 3 and 4 present the results for an example network composed of two primary resources and one secondary resource. The parameters of the system under investigation

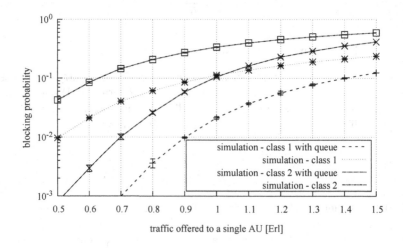

Fig. 2. Blocking probability in the primary resources no. 1

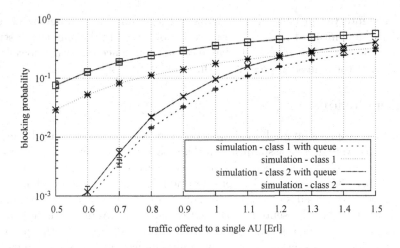

Fig. 3. Blocking probability in the primary resources no. 2

were as follows: $r = 2$, $M = 3$, $V_0 = 15$; $V_1 = 20$, $Q_1 = 10$, $t_1 = 1$, $t_3 = 3$; $V_2 = 10$, $Q_2 = 10$, $t_1 = 1$, $t_2 = 2$.

The impact of using the queue in primary resources of the presented system can be observed in Figs. 2 and 3. The calls blocked in the primary resources are then offered to the alternative resources. The comparison of the simulation and analytical results of the blocking probability in the alternative resources can be further observed in Fig. 4. The data obtained on the basis of the simulation study are presented as points with the confidence intervals calculated after the t-Student distribution (with 95-percent confidence level) for 5 series with 10000 calls each (the classes with the lowest call intensity). The obtained value of the

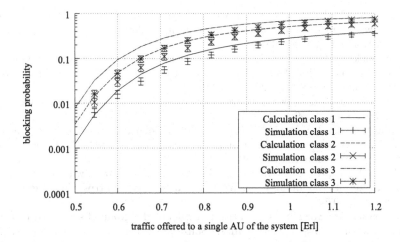

Fig. 4. Blocking probability in the alternative resources

confidence interval for each of the results of the simulation is lower by at least
an order of magnitude than the average value obtained in the simulation study.
In a large number of instances, the value of the confidence interval is lower that
the height of the symbol representing the simulation result.

5   Conclusions

This article proposes an analytical model of a multiservice hierarchical telecommunications system. The assumption in the model is that queues, that allow the
blocking probability to be decreased, are introduced in the primary resources of
the traffic overflow system under consideration. To validate the developed analytical model, the results of the calculations are compared with the simulation
data. Both the data presented in the article and the results gathered by the
authors as a result of relevant comparative study indicate that the proposed
model provides high accuracy in calculations. Further studies will be expanded
to include other types of offered traffic and a possibility of introducing queues
also in alternative resources.

Acknowledgements. This paper was developed as a result of the research project
2016/23/B/ST7/03925 entitled "Modelling and service quality evaluation of Internet-
based services" funded by the National Science Centre.

References

1. Bonald, T., Massoulié, L., Proutière, A., Virtamo, J.: A queueing analysis of max-
min fairness, proportional fairness and balanced fairness. Queueing Syst. **53**(1),
65–84 (2006). https://doi.org/10.1007/s11134-006-7587-7
2. Bonald, T., Roberts, J.W.: Internet and the Erlang formula. ACM SIGCOMM
Comput. Commun. Rev. **42**(1), 23–30 (2012). https://doi.org/10.1145/2096149.
2096153
3. Brune, G.: On delay and loss in a switching system for voice and data with internal
overflow. In: Proceedings of 11th International Teletraffic Congress, pp. 2.1–2.7.
North-Holland, Kyoto (1985)
4. Delbrouck, L.: On the steady-state distribution in a service facility carrying mix-
tures of traffic with different peakedness factors and capacity requirements. IEEE
Trans. Commun. **31**(11), 1209–1211 (1983)
5. Fredericks, A.: Congestion in blocking systems - a simple approximation technique.
Bell Syst. Tech. J. **59**(6), 805–827 (1980)
6. Glabowski, M., Kaliszan, A., Stasiak, M.: Modelling overflow systems with distributed secondary resources. Comput. Netw. **108**, 171–183 (2016). https://doi.
org/10.1016/j.comnet.2016.08.015
7. Glabowski, M., Kubasik, K., Stasiak, M.: Modeling of systems with overflow multirate traffic. Telecommun. Syst. **37**(1–3), 85–96 (2008). https://doi.org/10.1007/
s11235-008-9070-8
8. Głąbowski, M., Stasiak, M.: Multi-rate model of the group of separated transmission links of various capacities. In: de Souza, J.N., Dini, P., Lorenz, P. (eds.) ICT
2004. LNCS, vol. 3124, pp. 1101–1106. Springer, Heidelberg (2004). https://doi.
org/10.1007/978-3-540-27824-5_143

9. Haddad, J.P., Mazumdar, R.R.: Congestion in large balanced multirate networks. Queueing Syst. **74**(2), 333–368 (2013). https://doi.org/10.1007/s11134-012-9322-x
10. Hanczewski, S., Stasiak, M., Weissenberg, J.: A queueing model of a multi-service system with state-dependent distribution of resources for each class of calls. IEICE Trans. Commun. **E97–B**(8), 1592–1605 (2014)
11. Heffes, H.: Analysis of first-come first-served queuing systems with peaked inputs. Bell Syst. Tech. J. **52**(7), 1215–1228 (1973). https://doi.org/10.1002/j.1538-7305.1973.tb02014.x
12. Huang, Q., Ko, K.T., Iversen, V.B.: Approximation of loss calculation for hierarchical networks with multiservice overflows. IEEE Trans. Commun. **56**(3), 466–473 (2008)
13. Iversen, V.: Teletraffic engineering handbook. Technical report, Technical University of Denmark, Lyngby (2010)
14. Kaufman, J.S., Rege, K.M.: Blocking in a shared resource environment with batched Poisson arrival processes. J. Perform. Eval. **24**(4), 249–263 (1996). https://doi.org/10.1016/0166-5316(94)00029-8
15. Kaufman, L., Seery, J.B., Morrison, J.A.: Overflow models for dimension PBX feature packages. Bell Syst. Tech. J. **60**(5), 661–676 (1981). https://doi.org/10.1002/j.1538-7305.1981.tb00255.x
16. Kuczura, A.: The interrupted Poisson process as an overflow process. Bell Syst. Tech. J. **52**(3), 437–448 (1973). https://doi.org/10.1002/j.1538-7305.1973.tb01971.x
17. Lagrange, X., Godlewski, P.: Performance of a hierarchical cellular network with mobility-dependent hand-over strategies. In: Proceedings of Vehicular Technology Conference - VTC, pp. 1868–1872. IEEE, Atlanta, April 1996. https://doi.org/10.1109/VETEC.1996.504082
18. Matsumoto, J., Watanabe, Y.: Individual traffic characteristics queueing systems with multiple poisson and overflow inputs. IEEE Trans. Commun. **33**(1), 1–9 (1985). https://doi.org/10.1109/TCOM.1985.1096202
19. Meier-Hellstern, K.S.: Parcel overflows in queues with multiple inputs. In: Proceedings of 12th International Teletraffic Congress, pp. 3.1–3.8. North-Holland, Torino (1988)
20. Morrison, J.A.: Analysis of some overflow problems with queuing. Bell Syst. Tech. J. **59**(8), 1427–1462 (1980). https://doi.org/10.1002/j.1538-7305.1980.tb03373.x
21. Rácz, S., Gerő, B.P., Fodor, G.: Flow level performance analysis of a multi-service system supporting elastic and adaptive services. Perform. Eval. **49**(1–4), 451–469 (2002). https://doi.org/10.1016/S0166-5316(02)00115-3
22. Stamatelos, G.M., Koukoulidis, V.N.: Reservation-based bandwidth allocation in a radio ATM network. IEEE/ACM Trans. Netw. **5**(3), 420–428 (1997). https://doi.org/10.1109/90.611106
23. Stasiak, M.: Queuing systems for the internet. IEICE Trans. Commun. **E99–B**(6), 1224–1242 (2016)
24. Stasiak, M., Glabowski, M., Wiśniewski, A., Zwierzykowski, P.: Modeling and Dimensioning of Mobile Networks. Wiley, Hoboken (2011)
25. Wang, M., Li, S., Wong, E., Zukerman, M.: Performance analysis of circuit switched multi-service multi-rate networks with alternative routing. J. Lightwave Technol. **32**(2), 179–200 (2014). https://doi.org/10.1109/JLT.2013.2289925
26. Wilkinson, R.I.: Theories of toll traffic engineering in the USA. Bell Syst. Tech. J. **40**, 421–514 (1956)

Exploring YouTube's CDN Heterogeneity

Anh-Tuan Nguyen[1], Olivier Fourmaux[1(✉)], and Christophe Deleuze[2]

[1] Sorbonne Université, CNRS, LIP6, 75005 Paris, France
{anh-tuan.nguyen,olivier.fourmaux}@lip6.fr
[2] Univ. Grenoble Alpes, Grenoble INP
(Institute of Engineering Univ. Grenoble Alpes), LCIS,
26000 Valence, France
christophe.deleuze@lcis.grenoble-inp.fr

Abstract. In this paper, we set up measurements and make performance and geographic analysis of YouTube CDN video platform. We use large distributed testbeds, like PlatnetLab and EdgeNet, to grasp the heterogeneity of client situations. Those facilities can work as real clients without any simulation. From these infrastructures, we generate numerous requests to YouTube video servers. Using a reduced initial set of nodes in different geographic location, we continuously measure information related to YouTube homepage websites and video servers, and calculate the latency from clients to cache servers. We also look at the geographical location of YouTube servers. This enables a better understanding of cache mapping strategy and draws the map of the system. Our first result focus on distance between users and data centers before studying dynamic aspect of the system. The information we collect can be of interest to e.g. ISP network operators who need to improve their network architecture to minimize costs and optimize quality for the user.

Keywords: CDN · Measurement · YouTube · PlanetLab · EdgeNet

1 Introduction

The communication model of World Wide Web was initially designed with the content located and served from a unique server host. Users who want to access these content interact with the client that will generate requests through the Internet to the server. Time to transmit content may be long. Today, the classical structure of the network, especially for the web, has dramatically changed. Internet infrastructure and bandwidth needs have strongly increased, and the types of access have multiplied through the diversity of customers and wireless technologies. To follow this, the structure and content of the web has also changed. Multimedia content and video are growing fast and are used extensively. Websites also use automated settings with scripts from the server, adapting to very heterogeneous users for optimizing content and bandwidth. As a result, the number of users increased so much that websites cannot fit on a unique server

© ICST Institute for Computer Sciences, Social Informatics and Telecommunications Engineering 2019
Published by Springer Nature Switzerland AG 2019. All Rights Reserved
T. Q. Duong et al. (Eds.): Qshine 2018, LNICST 272, pp. 158–166, 2019.
https://doi.org/10.1007/978-3-030-14413-5_13

anymore and led to the development of new load balancing and content distribution systems. Initially, popular contents were simply cloned on many different servers located at many Data Centers in different locations around the world. A user needs to find and pick the closest mirror near him manually and can cause unbalanced situations. Nowadays, content distribution networks (CDNs) such as Akamai [1] or Amazon [2] rely on dynamic mechanisms. They set up dedicated DNS servers inside their networks, automatically analyzing the user's IP address to make adapted answer based on geographic proximity, resource costs, bandwidth availability, and other factors. Content from the origin server is then replicated and stored in or near the Internet provider network (ISP). This infrastructure is not correlated with the routing system and difficult to measure, predict and manage. With the rise of video traffic, we were motivated to run distributed measurement platform to get more information on one of the most popular video content providers: YouTube. Its video delivery is provided by dedicated worldwide CDN and we initiate a measurement campaign to get more understanding about it.

In Sect. 2, we give a brief CDN description and make some emphases on YouTube CDN. In Sects. 3 and 4 we introduce the testbeds used and the measurement we are conducting. In Sect. 5 we provide our first analysis of the Youtube CDN. In Sect. 6, related work is presented. Then we conclude and give some perspective in Sect. 7.

2   Context

To solve the problem of large scale content delivery, *content distribution* architectures have been designed and deployed. A *content distribution network* [6] is made of (among other things) a large number of servers in which the content is made available, plus a *redirection* system whose role is to direct a client to a chosen server. Information about the location of the user, the current load on the servers and on different parts of the network need to be taken into account in order to select the "best" server, *i.e.* the one that will provide the best quality of experience for the user. Also, the redirection needs to be performed *transparently*, without the user being aware of it. Classic approaches are providing different answers to DNS requests depending on the client, encoding client localization in a provided URL. We can also find the use of IP anycast addressing.

Here's a brief description of YouTube content distribution architecture. The user points its browser to a URL such as:

`https://www.youtube.com/watch?v=videoID`

The `www.youtube.com` domain name is resolved to the IP address a *homepage server* and an HTTP request is sent. This homepage server queries a *mapping server* with the client IP address and videoID, that replies with a DNS name (such as `r3---sn-gxo5uxg-jqbe`). This name is used to build a URL that points to the video stream such as:

`https://r3---sn-gxo5uxg-jqbe.googlevideo.com/videoplayback?ei...`

The browser now resolves the domain name to an IP address for a *video server* and gets the video stream from there.

3 Testbed

YouTube CDN is a large-scale distributed system. We consider to building a distributed measurement facility. We need to measure on a large number of clients, then we can have a bigger map of one of the largest CDN system in the world. We decided to choose a stable testbed: PlanetLab; and a very new: EdgeNet. They provides extended access to the system resources: open the socket, send/receive IGMP packets... directly without any system emulation. It allows measurements to be made without any simulator or below another system, which can change the results. With full permission on a machine, we can run unmodified clients, and the result of the measurements are realistic.

PlanetLab [7–9] is a global research network that supports the experimentation of network services. Since the beginning of 2003, more than 1,000 researchers have used PlanetLab to develop new technologies for distributed storage, network mapping, peer-to-peer systems, etc. PlanetLab currently consists of 1353 potential nodes at 717 sites but some of them are no longer maintained. Planetlab is split in several portions and as our Lab, LIP6, run the PlanetLab Europe's control center in Paris and so we chose mostly European sites. Initial measurement started with 27 IP addresses of PlatnetLab Europe nodes spread across Europe in 21 different cities.

EdgeNet [10,12] is a distributed edge cloud, in the family of PlanetLab, GENI, JGN-X, and PlanetLab Europe. It is a modern distributed edge cloud, incorporating advances in Cloud technologies over the past few years. EdgeNet is based on industry-standard Cloud software, with Docker as containerization technology and Kubernetes as the node manager and deployment solution. It is an opt-in global Kubernetes cluster; once a user has authenticated with this portal and been approved, she will be able to use standard Kubernetes tools and technologies to deploy an application across the EdgeNet infrastructure.

4 Measurements

We have developed a "measurement client" written in Python to perform all the steps a real browser would do to display a video. The tool records detailed log of the process including the HTTP request and response messages, the URLs involved at each step, the DNS resolutions performed and the IP addresses found, their geolocation, as well as timestamps for all the important events.

As described in Sect. 2, the host name www.youtube.com can be resolved to a number of different *homepage servers*. By repeating requests, we are able to find IP addresses of (some of) these servers. Our tool tries to behave like a normal client, it has a list of video content and manage pause between requests (pause length is chosen randomly in the interval [20, 1200] seconds).

We also knew that the HTML page returned by the homepage server contains URLs for the video themselves, in JavaScript snippets embedded in the page. By parsing the HTML page and the JavaScript snippets, the tool can extract the domain name for the video server (as r4---sn-gxo5uxg-jqbe.

googlevideo.com) and resolve it to find the associated IP address. It also queries a geolocation service and records geolocation information for each IP address found (for both video and homepage servers).

Finally, as soon as it has the result of a DNS resolution, the tool measures the round trip time with the server (with the ping utility) and records minimum, maximum and average latency.

5   Results

In the initial phase, we have run clients on 25 nodes of PlatnetLab and 14 nodes of EdgeNet platform, using network of 31 ISPs and covering total 34 cities, in which 12 European countries, 2 countries in North America spanning a total of 10 time zones. We plan to deploy our tool on many more nodes in the future, based on the results obtained from this initial campaign. With 80.000 queries, we collected 356 different IP addresses for the **homepage servers location**. Geolocation information shows that these servers are located in 28 cities of 11 countries. We found that 315 IP addresses are in the United States and Canada and 41 are in Europe. Considering the queries, 73% of the responses came from 20 data centers at different cities in the United States and Canada, and 27% from servers in other countries. Of these, 52% of responses were sent from the client's country. For each request sent to the YouTube website, we observed an average number of 86 HTTP requests and an average amount of 3 megabytes of data transferred. Most of requests finished in less than 6 seconds. In our result, some web servers have an IP belonging to a network supposed in the US but the delay is smaller. There is little mismatch between distance and delay for web servers (especially for PlanetLab nodes), because YouTube has a worldwide network it can allocate its addresses dynamically [15].

Concerning the **video servers location**, we found they are present in 34 cities in all of the 12 considered countries. There are 1173 different IP addresses in our measurement, so each location hosts several video servers. Of these addresses, 1112 are geo-located in the United States and Canada and 61 in Europe. However, contrary to what we saw for homepage servers, most of responses were received from within Europe: 53,8% of responses were sent from an European country, and 46,2% sent from the United States. More importantly, 75,5% of the responses were sent from the client's country. This suggests that geographic distance with the client is (or is strongly correlated with) one of the top criteria YouTube uses to select the video server for a given client.

A response HTML page of YouTube homepage is rather complex. Figure 1 shows the loading of such a response. Each line in this figure is a TCP connection, with purple color in a row is SSL initial connection, gray is stalled, green is time to first byte and blue is actual content loading. Blue vertical line is *DOM-ContentLoaded*, indicating when the initial markup of a page has been parsed and red vertical line is *loaded*, indicating when a page has been fully loaded. This map shows many static components that are mentioned in the following as images, js, css files... After analyzing all URLs from the HTML response of

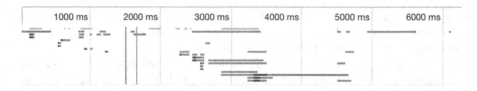

Fig. 1. Loading timeline map of a YouTube website connection. (Color figure online)

YouTube's server, we classify them into four categories: Homepage, Video cache host names, Static cache host names and unknown type.

The host name `www.youtube.com` is common to all the **homepage servers**. Each DNS resolution for this domain brings the IP address of one of the homepage servers, depending on load balancing of CDN, is serving for location of clients. By using a very strong DNS system inside their network, which is giving us different IP addresses of the same domain, they can redirect us to the most appropriate caching server based on their server selection strategy and load balancing at the connect moment.

A video can be served from many different **video cache servers**. We recorded video cache domains in form:

$$r\alpha\text{---}sn\text{-}\beta\text{-}\gamma.\texttt{googlevideo.com}$$

In which, α is 1–20, β and γ is alphabetic characters and digits. There is a large number of such video cache domains (we currently found 122), each being mapped to a number of different IPs. This introduces a large amount of flexibility and dynamicity, which can be used to redirect client requests to a video caching server selected to provide the best quality of service at a given time. As an example, we show in Table 1 some of the IPs we found for a precise video hostname (we found 30 different IPs for this name). After receiving our request, the homepage server analyzed and gave us a hostname, client received this hostname, resolved IP and make a connection to the server with this IP.

Inside HTML pages, other domains are visibly dedicated to serving **static content** such as JavaScript code, style sheets and images. Most of these are sub-domains of the `ytimg.com` domain, such as `s.ytimg.com`, `i.ytimg.com`, and `i9.ytimg.com`. The different kinds of static content are classified in the path part of the URL, as sub-directories: `./yts/jsbin/` for JavaScript codes, `./yts/cssbin/` for style sheets and `./yts/img` for images.

Several other types of domain names in response HTML can be observed, but they are not used frequently and their use was not investigated further.

We also study the **latency** and as expected, the average latency time collected for video server requests is smaller than for the homepage server requests. General, the maximum of latency of homepage and video servers is not much different: 196.145 ms for homepage servers and 185.6 ms for video servers. The minimum latency is very low (below 1 ms) for both kinds of servers, suggesting some of our nodes were very close to some YouTube servers. The average latency is much better for video servers (10.47 ms, versus 20.22 ms for homepage servers). This is in line with the result of our geolocation methodology that

Table 1. Some of the IPs & Geolocation of a video cache domain name
`r2---sn-gxo5uxg-jqbe.googlevideo.com`

Resolved IP	City	Region	Country
172.217.133.233	Ashburn	Virginia	United States
173.194.164.39	Ashburn	Virginia	United States
173.194.187.9	Ashburn	Virginia	United States
173.194.190.106	Ashburn	Virginia	United States
173.194.190.170	Ashburn	Virginia	United States
74.125.105.122	Mountain View	California	United States
74.125.105.124	Mountain View	California	United States
172.217.130.233	Newark	New Jersey	United States
209.85.230.156	Newark	New Jersey	United States
173.194.5.8	Willard	North Carolina	United States
193.51.224.141	Paris	Île-de-France	United States
172.217.133.10	Ashburn	Virginia	United States

servers in the same country with clients have lower latency. We repeated the
latency measurement as well as clients replay the video or select a new video.
The collected result included some variation latency of requests from the same
client come to the same server at a different time. From the plots, we can see
the relation between latency and the distance (Table 2).

Table 2. Percentage of number of requests with latency

Server/Latency less than	1 ms	5 ms	10 ms	20 ms	35 ms	50 ms	100 ms
Homepage	2.27%	31.1%	46.3%	67.15%	80.23%	85.55%	98.77%
Video	14.9%	57.1%	77.2%	89.6%	91.9%	95.3%	99.9%

With the result obtained, most of video request came from Europe will be
response from a video cache server inside Europe, almost that response sent from
the same country with client. So YouTube server selection strategy is seems effective, it not only gave client the nearest geography server but also the videos are
being delivered from a preferred data center. In Fig. 2, we can see the distribution of Video Cache Data Centers spread in Europe and concentrate at the
east side of USA. Opposite to Video Cache Data Center location, the location
of Homepage Cache Data Centers spread from west to east of USA but more
concentrate at the center of Europe. The number of Video Cache Data Center
is quiet high than homepage (Fig. 3).

Fig. 2. Distribution of video cache server data center.

Fig. 3. Distribution of homepage cache server data center.

6   Related Work

There already exists in a number of publications related to YouTube, most of them mainly focus on user behaviors or the system performance. Some other studies make measurement focus on Quality of Service [13]. They found that cache server selection is highly ISP-specific and that geographical proximity is not the primary criterion. In this paper they found YouTube uses HTTP via TCP to deliver the video, and now on UDP with Quic [14]. On the other hand, papers focusing on network-related issues seem less plentiful. A paper focus on DNS, domain name, and latency and throughput measurements are available in [16] for YouTube but a long time ago (2012), so the architecture may have changed. Another paper [17] focus on CDN selection strategy employed by YouTube, and they conclude that at least 10% request will be redirected to a non-preferred server, and the reasons are load balancing, variation DNS servers, load of hot spots due do popular content and unpopular content on given data centers.

7 Conclusion and Perspectives

In this paper we try to get a better understanding of the current CDN spreading structure of YouTube and how video requests are served by their data centers. Our work has been based on datasets collected from the edge of dozens different networks, ISP, located in different countries. Our measurement indicates that the YouTube infrastructure has been changed compared to the one previously analyzed in the publications. In the present system, most of YouTube response are based on geography distance, and they are focusing on video cache servers much more than before. The number of video servers deployed on ISP much more than homepage server. In which, the number of IPs in the United States much higher than in Europe absolutely. Because of this, we need to investigate deeper into their infrastructure to understand how do they do that, and gain their selection strategy. The selection strategy is a very important research, it can give us a big picture of routing and load balancing. We are continuously upgrading our measurement system and expand it geographically, to collect servers at a wider scale. By using the new testbed platform EdgeNet, we will be able to collect more data from the United States and other regions, thereby reinforces the value obtained. This work we have limitations. The main one is the limited reach of clients. All measurements are done by a "client view" which is receiving and analyzing the data sent from YouTube only, it's just an "edge view" of the system. Another point is related to the content that may impact on the server location and that we didn't take into account at this time.

References

1. Akamai. https://www.akamai.com. Accessed 04 Oct 2018
2. Amazon CloudFront. https://aws.amazon.com/cloudfront/. Accessed 04 Oct 2018
3. Azure CDN. https://azure.microsoft.com/en-us/services/cdn/. Accessed 04 Oct 2018
4. Cloudflare CDN. https://www.cloudflare.com/cdn/. Accessed 04 Oct 2018
5. YouTube. https://www.youtube.com/. Accessed 04 Oct 2018
6. Nygren, E., Sitaraman, R.K., Sun, J.: The akamai network: a platform for high-performance internet applications. ACM SIGOPS Operating Syst. Rev. **44**(3), 2–19 (2010)
7. PlanetLab. https://www.planet-lab.org/. Accessed 04 Oct 2018
8. PlanetLab Europe. https://planet-lab.eu/. Access 04 Oct 2018
9. Bavier, A., et al.: Operating system support for planetary-scale network services. In: USENIX Symposium on Networked Systems Design and Implementation, San Francisco (2004)
10. EdgeNet. https://edge-net.org/. Accessed 04 Oct 2018
11. GitHub source code. https://github.com/anh-tuan/YouTube-CDN-measurement. Accessed 14 Nov 2018
12. Cappos, J., Hemmings, M., McGeer, R., Rafetseder, A., Ricart, G.: EdgeNet: a global cloud that spreads by local action. In: Proceedings of ACM/IEEE Symposium on Edge Computing, Bellevue (2018)

13. Hoßfeld, T., Schatz, R., Biersack, E., Plissonneau, L.: Internet video delivery in youtube: from traffic measurements to quality of experience. In: Biersack, E., Callegari, C., Matijasevic, M. (eds.) Data Traffic Monitoring and Analysis. LNCS, vol. 7754, pp. 264–301. Springer, Heidelberg (2013). https://doi.org/10.1007/978-3-642-36784-7_11
14. Langley, A., et al.: The QUIC transport protocol: design and internet-scale deployment. In: Proceedings of the SIGCOMM 2017, New York, pp. 183–196 (2017)
15. de Vries, W.B., van Rijswijk-Deij, R., de Boer, P.T., Pras, A.: Passive observations of a large DNS service: 2.5 years in the life of Google. In: Proceedings of Network Traffic Measurement and Analysis Conference (TMA 2018), Vienna, Austria (2018)
16. Adhikari, V. K., Jain, S., Chen, Y., Zhang, Z.L.: Vivisecting YouTube: an active measurement study. In: Proceedings of IEEE INFOCOM, Orlando, pp. 2521–2525 (2012)
17. Torres, R., Finamore, A., Kim, J.R., Mellia, M., Munafo, M.M., Rao, S.: Dissecting video server selection strategies in the YouTube CDN. In: 31st International Conference on Distributed Computing Systems, Minneapolis, pp. 248–257 (2011)

Author Index

Printed in the United States
By Bookmasters